Encyclopaedia of
Nanotechnology

Encyclopaedia of **Nanotechnology**

Vol. 9 : Construction Engineering

Dr. Rita Khare

AGROTECH PRESS

JAIPUR ■ NEW DELHI

AGROTECH PRESS

Regd. Off. "Bony Residency", Gate No. 2
Opp. Tilak Public School, Vishweswariya Nagar,
Gopalpura Road, Jaipur - 302018, India
Ph: 0141 - 2761280, Telefax: 2761381
E-mail: agrotechpress@gmail.com

Delhi Office: 102, 1st Floor, Satyam House,
4327/3, Ansari Road, Darya Ganj,
New Delhi - 110002, India
Ph: 011-45652440
visit us: www.agrotechpress.com

Distributed by :
Keya International Press LLP
E-mail: keyainternationalpressllp@gmail.com

Published in India, 2024

ISBN : 9789384568566 (10 Vol. Set)

Typeset by Agrotech Press

Printed and bound in India by Replika Press Pvt. Ltd.

Contents

Preface

Construction as a science or technology has undergone great changes over its history. Civil engineering profession has delivered infrastructures of various scales and complexities to different generations, as far as human history can stretch. Over this period, the demands and expectations from the profession have increased. Problems stemmed from our living and consumption patterns vis-a-vis interaction with the wider ecosystem. Sustainable development as a concept has evolved as a multidimensional, straddling across socio-economic, ecological, and technical perspective. The civil engineering and construction sector heralded a paradigm shift from traditional infrastructure procurement and delivery with the focus now on sustainable infrastructure systems in the last century. However, appropriate technologies are needed to transform existing construction processes in product development. This text discusses an innovative technology that offers solution to the challenges facing civil engineering in the 21st century. Nanotechnology is an underpinning technology that addresses some of the infrastructure sustainability challenges. The volume discusses the application of nanotechnology and nanomaterials for sustainable infrastructure delivery, and charts directions for further research.

The role of nanotechnology in conceiving of innovative infrastructure systems has potential to revolutionize the civil engineering practice. Nanotechnology is a field that is dominated by developments in basic physics and chemistry research, where phenomena on atomic and molecular level are used to provide materials and structures that perform tasks that are not possible using the materials in their typical macroscopic form. It is rather an extension of the sciences and technologies already developed for many years ,to examine the nature of our world at an ever smaller scale. Nanotechnology is the use of very small particles of material. Nanotechnology is a science concerned with the design, construction and utilization of functional structures with at least one characteristic dimension measured in nanometres. The size of the particles is very important because at the length scale of the nanometer, 10-9m, the properties of the material actually become affected. Nanotechnology initially developed in the fields of physics and chemistry, and most fundamental developments still occur in these fields. Nanotechnology also needs to be applied in areas such as the engineering field. Obviously, the application of nanotechnology to science and engineering has increased in other fields over the years. One area which is one of the most active research areas in the field of nanotechnology is civil engineering. This book presents a broad overview of the application of nanotechnology in the civil engineering.

This book presents an overview of nanotechnology and civil engineering in general and their interdisciplinary approach. Further the Nano-materials and their applications in civil engineering field across its different sections are exemplified. The construction business will inevitably be a beneficiary of this nanotechnology. In fact it already is in the fields of concrete, steel and glass, and many more. Concrete is stronger, more durable and more easily placed; steel is made tougher ; glass is self-cleaning. Paints are made more insulating andwater repelling. The properties like self-sensing, self-rehabilitation, self-structural health monitoring, self-vibration damping, self-cleaning and self-healing are studied. Following this the analysis were carried out in ductile structural composites along with its enhanced properties, low maintenance coatings, better properties of cementitious materials, reduction of the thermal transfer rate of fire retardant and insulation, various nanosensors, smart materials, intelligent structure technology etc. And also the possible consequences regarding its practicality and the most challenging economic factors concerned are discussed briefly. Finally its future trend and potential towards more economical, sustainable & eco-friendly infrastructure having longer durability are elucidated. The book also discusses developments in nano-engineering and nano-modification of construction materials to improve the bulk properties, development of sustainable, intelligent, and smart concrete materials through the integration of nanotechnology based self-sensing and self-powered materials and cyber infrastructure technologies, review of nanotechnology applications in pavement engineering, development of novel, cost-effective, high-performance and long-lasting concrete products and processes through nanotechnology-based innovative processing of cement and cement paste, and advanced nanoscience modeling, visualization, and measurement systems for characterizing and testing civil infrastructure materials at the nano-scale.

This volume aims to discuss various concepts behind nanotechnology, implementation of nanotechnology in Civil Engineering field. Research in nanotechnology that is related to construction is still in its infancy; however, this paper has demonstrated the main benefits and barriers that allow the effect of nanotechnology on construction to be defined. In this a broad overview of the potential application of various nanotechnology developments in the construction engineering field is discussed, and the potential for further basic research that may lead to improved systems is evaluated.

Dr. Rita Khare

1

Introduction

Construction engineering is a professional discipline that deals with the designing, planning, construction, and management of infrastructures such as highways, bridges, airports, railroads, buildings, dams, and utilities. These Engineers are unique such that they are a cross between civil engineers and construction managers. Construction engineers learn the designing aspect much like civil engineers and construction site management functions much like construction managers.

The primary difference between a construction engineer and a construction manager is that the construction engineer has the ability to sit for the Professional Engineer license (PE) whereas a construction manager cannot. At the educational level, construction managers are not as focused on design work as they are on construction procedures, methods, and people management. Their primary concern is to deliver a project on time, within budget, and of the desired quality.

The difference between a construction engineer and civil engineer is only at the educational level as both disciplines are able to sit for the PE exam giving them the same title of engineer. Civil engineering students concentrate more on the design work, gearing them toward a career as a design professional. This essentially requires them to take a multitude of design courses. Construction engineering students take design courses as well as construction management courses. This allows them to understand both the design functions as well as the building requirements needed to design and build today's infrastructures.

The role of nanotechnology in conceiving of innovative infrastructure systems has potential to revolutionize the civil engineering practice. As people involved in construction, we are very familiar with the concept of getting raw materials, bringing them together in an organised way and then putting them together into a recognisable form. The finished product is a passive machine. It works and slowly decays as it is used and abused by the environment and the owners of the project.

Construction then is definitely not a new science or technology and yet it has undergone great changes over its history. In the same vein, nanotechnology is not a new science and it is not a new technology either. It is rather an extension of the sciences and technologies that have already been in development for many years. The size of the particles is the critical factor. At the Nano scale, material properties are altered from that of larger scales. It is these "Nano-effects", however, that ultimately determine all the properties that we are familiar with at our "macro-scale" and this is where the power of nanotechnology comes in –if we can manipulate elements at the Nano scale we can affect the macro-properties and produce significantly new materials and processes.

The construction industry was the only industry to identify nanotechnology as a promising emerging technology in the UK Delphi survey in the early 1990s. The importance of nanotechnology was also highlighted in foresight reports of Swedish and UK construction. Furthermore, ready mix concrete and concrete products were identified as among the top 40 industrial sectors likely to be influenced by nanotechnology in 10-15 years.

However, construction has lagged behind other industrial sectors where nanotechnology R&D has attracted significant interest and investment from large industrial corporations and venture capitalists. Recognising the huge potential and importance of nanotechnology to the construction industry, the European Commission in late 2002 approved funding for the Growth Project GMA1-2002-72160 "NANOCONEX" -Towards the setting up of a network of excellence in nanotechnology in construction.

Applications of Nanotechnology in Civil Engineering

Nanotechnology can be used for design and construction processes in many areas since nanotechnology generated products have many unique characteristics. These characteristics can, again, significantly fix current construction problems, and may change the requirement and organisation of construction process.

Some of its applications are examined in detail below:

Concrete is one of the most common and widely used construction materials.

Nanotechnology is widely used in studying its properties like hydration reaction, alkali silicate reaction (ASR) and fly ash reactivity. Alkali silicate reaction is caused due to alkali content of cement and silica present in reactive aggregates. The use of pozzolona in the concrete mix as a partial cement replacement can reduce the likelihood of ASR occurring as they reduce the alkalinity of a pore fluid.

Fly ash not only improves concrete durability and strength but also sustainable factors by reducing the requirement for cement, however, the curing process of such concrete is slowed down due to the addition of fly ash and early stage strength is also low in comparison to normal concrete. Addition of Nano-silica leads to the densifying of the micro and nanostructure resulting in improved mechanical properties. With the addition of nano-silica part of the cement is replaced but the density and strength of the fly-ash concrete improves particularly in the early stages. For concrete containing large volume fly ash, at early age it can improve pore size distribution by filling the pores between large fly ash and cement particles at Nano scale. The dispersion/slurry of amorphous nano-SiO2 is used to improve segregation resistance for self-compacting concrete.

The addition of small amount of carbon nanotube (1%) by weight could increase both compressive and flexural strength. This can also improve the mechanical properties of samples consisting of the main portland cement phase and water. Addition of 1% of Oxidized multi-walled nanotubes (MWNT's) show the best improvements both in compressive strength (+ 25 N/mm2) and flexural strength (+ 8 N/mm2) compared to the reference samples without the reinforcement.

Cracking is a major concern for many structures. University of Illinois Urbana-Champaign is working on healing polymers, which include a microencapsulated healing agent and a catalytic chemical trigger. When the microcapsules are broken by a crack, the healing agent is released into the crack and contact with the catalyst. The polymerization happens and bond the crack faces. The self-healing polymer could be especially applicable to fix the micro cracking in bridge piers and columns. But it requires costly epoxy injection.

Research has shown that an anaerobic (one that does not require oxygen) microorganism in corporated into concrete mixing water results in a 25% increase in 28-day strength. The Shewanella microorganism was used at a concentration of 10^5cells/ml and nanoscale observation revealed that there was a deposition of sand-cement matrix on its surface. This led to the growth of filler material within the pores of the cement sand matrix and resulted in increased strength.

Finally, fibre wrapping of concrete is quite common today for increasing the strength of pre-existing concrete structural elements. An advancement in the procedure involves the use of a fibre sheet (matrix) containing nano-silica particles and hardeners. These nanoparticles penetrate and close small cracks on the concrete surface and, in strengthening applications, the matrices form a strong bond between the surface of the concrete and the fibre reinforcement. It is evident from the Figure depicted earlier, that the SCCNFC (self- consolidating carbon Nano fibre concrete) column failed at higher loads and with larger deflection than the SCRC (steel confined reinforced concrete) column. Additionally, the SCCNFC column was much stiffer than the SCRC column and exhibited higher energy dissipation. SCCNFC can also be used as a type of self-Structural Health Monitoring system.

Structural Composites Steel is a major construction material. FHWA together with American Iron and Steel Institute and the U.S. Navy developed new, low carbon, high-performance steel (HPS) for bridges in 1992 with higher corrosion-resistance and weld ability by incorporating copper nanoparticles from at the steel grain boundaries.

Sandvik NanoflexTM is new stainless steel developed by Sandvik Nanoflex Materials Technology. Due to its high performance, it is suitable for application which requires lightweight and rigid designs. Its good corrosion, formability and wear resistance can keep life-cycle costs low.

MMFX2 is nanostructure -modified steel, produced by MFX Steel Corp, USA.

Compared with the conventional steel, it has a fundamentally different microstructure-laminated lath structure resembling "plywood" as shown in Figure. Due to the modified nanostructure, MMFX steel has superior mechanical properties, e.g. higher strength, ductility and fatigue resistance, over other high -strength steels. These material properties can lead to longer service life in corrosive environments and lower construction costs. The MMFX2 steel has similar corrosion resistance to that of stainless steel, but at a much lower cost. So far, the MMFX steel has gained certification for use in general construction throughout the US.

Carbon nanotubes are over 100 times stronger than steel and only one-sixth of the weight in addition to its high thermal and electrical conductivities. A CNT composite has recently been reported to be six times stronger than conventional carbon fibre composites. Additionally, unlike carbon fibres which fracture easily under compression, the nanotubes are much more flexible and can be compressed without fracturing. CNT composite reinforced structures have a 50 to 150-fold increase in tensile strength, compared with conventional steel-reinforced structures.

Coatings

The coatings incorporate certain Nano particles or Nano layers have been developed for certain purpose including: protective or anti-corrosion coatings for components; self-cleaning, thermal control, energy saving, anti-reflection coatings for glass/windows; easy-to-clean, antibacterial coatings for work surfaces; and more durable paints and anti-graffiti coating for buildings and structures. For example:

Self-cleaning windows have been developed and marketed by Pilkington, St. Gobain Co., and others. This coating works in two stages. First, using a ' photocatalytic' process, nanosized TiO2 particles in the coating react with ultra-violet rays from natural daylight to break down and disintegrate organic dirt. Secondly, the surface coating is hydrophilic, which lets rainwater spread evenly over the surface and 'sheet' down the glass to wash the loosened dirt away. It can therefore reduce airborne pollutants when applied to outdoor surfaces.

Coating of 7000 m2 of road surface with such a material in Milan in 2002 has led to a 60% reduction in nitrogen oxides concentration at street level. Research has also demonstrated that bimetallic Nano particles, such as Fe/Pd, Fe/Ag, or Zn/Pd, can serve as potent reductants and catalysts for a large variety of environmental contaminants.

Another approach to create self-cleaning surface coating has been the development of 'Lotus Spray' products by BASF, based on ideas of replicating the spotless lotus leaves by incorporating silica and alumina nanoparticles and hydrophobic polymers. The product offers 20 times more water-repellent property than a smooth, wax coating. With its applications in the construction industry, the company aims to develop a product that will retain its lotus effect even after an abrasion with sandpaper.

Special coatings can also make the applied surface both hydrophobic and oleophobic at the same time. These could be used for anti-graffiti surfaces, carpets and protective clothing etc. Researchers in Mexico has successfully developed a new type of anti-graffiti paint DELETUM, by functionalising nanoparticles and polymers to form a coating repellent to water and oil at the same time.

As a result, the coated surface is non-stick or very easy to clean, and able to withstand repeated graffiti attacks.

Furthermore nanostructured coatings can be used to selectively reflect and transmit light in different wavebands. Research is focusing on smart and responsive materials able to sense and adapt to surroundings and change their appearance, such as whose colour changes as a function of temperature, and cladding which responds to heat and light to minimise energy use in buildings.

Glass Fire-protective glass is another application of nanotechnology. This is achieved by using a clear intumescent layer sandwiched between glass panels (an interlayer) formed of fumed silica (SiO2) nanoparticles which turns into a rigid and opaque fire shield when heated. The electrochromic coatings are being developed that react to changes in applied voltage by using a tungsten oxide layer; thereby becoming more opaque at the touch of a button. Because of the hydrophobic properties of TiO2, it can be applied in antifogging coatings or in self-cleaning windows. Nano-TiO2 coatings can also be applied to building exteriors to prevent sticking of pollutants, and thus reduce a facility's maintenance costs.

Nanosensors

Nanotechnology enabled sensors/devices which exhibit 'self-sensing' and 'self-actuating' capability also offer great potential for developing smart materials and structures.

The device used for air bags in cars is such an example. Nano and Micro electrical mechanical systems (NEMS & MEMS) sensors range from 10-9m to 10-5 m which could be embedded into the structure during the construction process. They can monitor and/or control the environment conditions (e.g. temperature, moisture, smoke, noise, etc.) and the materials/structure performance (e.g. stress, strain, vibration, cracking, corrosion etc.) during the structure's life. Smart aggregate, a low cost piezoceramic-based multi-functional device, has been applied to monitor early age concrete properties such as moisture, temperature, relative humidity and early age strength development. Also it can provide an early indication before a failure of the structure occurs. Thus the sensors are able to work as self-health monitoring system.

Cyrano Sciences has developed electronic noses based on an array of different polymer nanometre-thin film sensors. Siemens and Yorkshire Water are developing autonomous, disposable chips with built -in chemical sensors to monitor water quality and send pollution alerts by radio.

Bulk Insulating Materials

Nano Pore has developed bulk nanoporous silica compounds with embedded organic molecules which perform up to 10 times better than conventional insulating materials. The superior insulation haracteristics of these low densities, highly porous solids are due to the unique shape and small size (10-100 nm) of its large number of pores. So far, these new insulating compounds have been used in applications that require excellent thermal performance, optimum energy efficiency, or minimum insulation thickness.

Plastics

The carbon fibre reinforced plastics (CFRP) being light weight material does not exhibit good electrical properties. CNTs are among the stiffest and strongest fibres known, and can improve electrical conductivity and heat dissipation properties. At IFAM in Bremen, researchers employed plasma technology in order to transfer their properties to CFRPs since these micro-or nanoparticles must be highly homogeneous, and made such as to readily bound to the polymer.

This can be pressed onto an electronic component so heat is dissipated directly.

Plastic Solar Cell

The most promising application in the areas of energy and environment leading to the sustainable building is the development of fuel cells and photovoltaic. In the last few years, considerable efforts have been made to develop plastic solar cells in Figure earlier, with a nanorod/polymer layer sandwiched between two electrodes. The middle layer, a mere 200 nm thick, is a jumble of nanorods embedded in the semiconducting polymer., much simpler and cheaper to produce than that of conventional silicon semiconductor solar cells.

Nanotechnology in Civil Engineering

Nanotechnology in Civil Engineering, Nanotechnology is an enabling technology that allows us to develop materials with improved or totally new properties". It is rather an extension of the sciences and technologies already developed for many years ,to examine the nature of our world at an ever smaller scale. Nanotechnology is the use of very small particles of material. A nanometre is a billionth of a metre.

The size of the particles is very important because at the length scale of the nanometre, 10-9m, the properties of the material actually become affected.

Civil Engineering is always restricted to cement, sand and stones. But it is something beyond that. It deals with innovations in each and every step of constructing a structure. This paper deals with a "NONATECHNOLOGICAL TREATMENTS" IN Civil Engineering.

Nanotechnology in Construction

The construction business will inevitably be a beneficiary of this nanotechnology. In fact it already is in the fields of concrete, steel and glass, and many more. Concrete is stronger, more durable and more easily placed; steel is made tougher ; glass is self-cleaning. Paints are made more insulating and water repelling.

Introduction to Nano Materials: Nano particle: It is defined as a particle with at least one dimension less than 200nm. It is quantum dots if they are small enough (typically sub 10nm) such that jumps in energy levels occur.

Nano composite : It is produced by adding Nano particle to a Bulk material in order to improve the bulk material properties.

Carbon Nano Tubes(CNT):

- They are cylindrical with nanometre diameter.
- They can be several millimetres in length.
- They have 5 times the Young's modulus and 8 times (theoretically 100 times) the strength of steel whilst being 1/6th the density.
- Thermal conduction is also very high along the tube axis

Titanium oxide:

- Titanium dioxide is a widely used white pigment.
- It can oxidize oxygen or organic materials, and so added to paints, cements, windows, tiles, or other products for sterilizing, deodorizing & anti-fouling properties
- When incorporated into outdoor building materials can substantially reduce concentrations of airborne pollutants.

Additionally, as TiO2 is exposed to UV light, it becomes increasingly hydrophilic, thus it can be used for anti-fogging coatings or selfcleaning windows.

Nanotechnology in Concrete: Concrete is a mixture of cement, sand (fine aggregate), coarse aggregate & water. As concrete is most usable material in construction industry it's been require to improve its quality.

The mechanical behavior of concrete materials depends on phenomena that occur on a micro and a nano scale.

Nano-concrete is defined as "A concrete made with Portland cement particles that are less than 500 Nano-metres as the cementing agent".

- Concrete is, after all, a macro-material strongly influenced by its nano-properties.
- NANO-SILICA: particle packing in concrete can be improved by using nano-silica which leads to a densifying of the micro & nanostructure resulting in improved mechanical properties
- Nano-silica addition to cement based materials can also control the degradation of the fundamental C-S-H (calcium-silicate hydrate) reaction of concrete caused by calcium leaching in water as well as block water penetration and therefore lead to improvements in durability.
- Related to improved particle packing, high energy milling of ordinary Portland cement (OPC) clinker and standard sand, produces a greater particle size diminution with respect to conventional OPC and, as a result, the compressive strength of the refined material is also 3 to 6 times higher.

TiO2 in Concrete:

- TiO2 is a white pigment and can be used as an excellent reflective coating.
- It is hydrophilic and therefore gives self cleaning properties to surfaces to which it is applied.
- The process by which this occurs is that rain water is attracted to the surface and forms sheets which collect the pollutants and dirt particles previously broken down and washes them off.
- The resulting concrete, already used in projects around the world, has a white colour that retains its whiteness very effectively unlike the stained buildings of the material's pioneering past.

CNTs in Concrete:

- The addition of small amounts (1% wt) of CNT's can improve the mechanical properties of samples consisting of the main Portland cement phase and water.
- Oxidized multi-walled Nano tubes (MWNT's) show the best improvements both in compressive strength (+ 25 N/mm2) and flexural strength (+ 8 N/mm2) compared to the samples without the reinforcement.
- A number of investigations have been carried out for developing smart concrete using carbon fibres.

Asphalt

Asphalt, also known as bitumen, is a sticky, black and highly viscous liquid or semi-solid form of petroleum. It may be found in natural deposits or may be a refined product; it is a substance classed as a pitch. Until the 20th century, the term asphaltum was also used.

The primary use (70%) of asphalt/bitumen is in road construction, where it is used as the glue or binder mixed with aggregate particles to create asphalt concrete. Its other main uses are for bituminous waterproofing products, including production of roofing felt and for sealing flat roofs.

The terms *asphalt* and *bitumen* are often used interchangeably to mean both natural and manufactured forms of the substance. In American English, asphalt (or asphalt cement) is the carefully refined residue from the distillation process of selected crude oils.

Outside the United States, the product is often called bitumen. Geological terminology often prefers the term *bitumen*. Common usage often refers to various forms of asphalt/bitumen as "tar", such as at the La Brea Tar Pits. Another term, mostly archaic, refers to asphalt/bitumen as "pitch".

Naturally occurring asphalt/bitumen is sometimes specified by the term "crude bitumen". Its viscosity is similar to that of cold molasses while the material obtained from the fractional distillation of crude oil [boiling at 525 °C (977 °F) is sometimes referred to as "refined bitumen".

Composition

The components of asphalt are classified into four classes of compounds:

- saturates, saturated hydrocarbons, the %saturates correlates with softening point of the material
- Naphthene aromatics, consisting of partially hydrogenated polycyclic aromatic compounds.
- Polar aromatics, consisting of high molecular weight phenols and carboxylic acids
- Asphaltenes, consisting of high molecular weight phenols and heterocyclic compounds

The naphthene aromatics and polar aromatics are typically the majority components. Additionally, most natural bitumens contain organosulfur compounds, resulting in an overall sulfur content of up to 4%. Nickel and vanadium are found in the <10 ppm level, as is typical of some petroleum.

The substance is soluble in carbon disulfide. It is commonly modelled as a colloid, with asphaltenes as the dispersed phase and maltenes as the continuous phase. and "it is almost impossible to separate and identify all the different molecules of asphalt, because the number of molecules with different chemical structure is extremely large".

Asphalt/bitumen can sometimes be confused with "tar", which is a similar black, thermoplastic material produced by the destructive distillation of coal.

During the early and mid-20th century when town gas was produced, tar was a readily available product and extensively used as the binder for road aggregates. The addition of tar to macadam roads led to the word tarmac, which is now used in common parlance to refer to road-making materials.

However, since the 1970s, when natural gas succeeded town gas, asphalt/bitumen has completely overtaken the use of tar in these applications. Other examples of this confusion include the La Brea Tar Pits and the Canadian tar sands. Pitch is another term mistakenly used at times to refer to asphalt/bitumen, as in Pitch Lake.

Occurrence

Figure: *Bituminous outcrop of the Puy de la Poix, Clermont-Ferrand, France*

The great majority of asphalt used commercially is obtained from petroleum. Nonetheless, large amounts of asphalt occur in concentrated form in nature. Naturally occurring deposits of asphalt/bitumen are formed from the remains of ancient, microscopic algae (diatoms) and other once-living things. These remains were deposited in the mud on the bottom of the ocean or lake where the organisms lived. Under the heat (above 50 °C) and pressure of burial deep in the earth, the remains were transformed into materials such as asphalt/bitumen, kerogen, or petroleum. Deposits at the La Brea Tar Pits are an example.

Natural deposits of asphalt/bitumen include lakes such as the Pitch Lake in Trinidad and Tobago and Lake Bermudez in Venezuela. Natural seeps of asphalt/bitumen occur in the La Brea Tar Pits and in the Dead Sea. Asphalt/bitumen also occurs as impregnated sandstones known as bituminous rock and the similar "tar sands" such as in Athabasca, Canada and Utah, USA. The Athabasca tar sands are located in the McMurray Formation, Alberta. This Formation is of early Cretaceous age, and is composed of numerous lenses of oil-bearing sand with up to 20% oil. Isotopic studies attribute the oil deposits to be about 110 Ma old. Heavy oil or bitumen deposits also occur in the Uinta Basin in Utah, USA. The Tar Sand Triangle deposit, for example, is roughly 6% bitumen.

Asphalt/bitumen occurs in hydrothermal veins. An example of this is within the Uinta Basin of Utah, USA, where there is a swarm of laterally and vertically extensive veins composed of a solid hydrocarbon termed Gilsonite. These veins formed by the polymerisation and solidification of hydrocarbons that were mobilized from the deeper oil shales of the Green River Formation during burial and diagenesis.

Asphalt/bitumen is similar to the organic matter in carbonaceous meteorites. However, detailed studies have shown these materials to be distinct.

History

Ancient Times: The use of asphalt/bitumen for waterproofing and as an adhesive dates at least to the fifth millennium B.C. in the early Indus community of Mehrgarh, where it was used to line the baskets in which they gathered crops.

In the ancient Middle East, the Sumerians used natural asphalt/bitumen deposits for mortar between bricks and stones, to cement parts of carvings, such as eyes, into place, for ship caulking, and for waterproofing. The Greek historian Herodotus said hot asphalt/bitumen was used as mortar in the walls of Babylon.

In some versions of the Book of Genesis in the Bible, the name of the substance used to bind the bricks of the Tower of Babel is translated as bitumen, while other translations use the word *pitch*. A one-kilometre tunnel beneath the river Euphrates at Babylon in the time of Queen Semiramis (ca. 800 B.C.) was reportedly constructed of burnt bricks covered with asphalt/bitumen as a waterproofing agent.

Asphalt/bitumen was used by ancient Egyptians to embalm mummies. The Persian word for asphalt is *moom*, which is related to the English word mummy. The Egyptians' primary source of asphalt/bitumen was the Dead Sea, which the Romans knew as *Palus Asphaltites* (Asphalt Lake).

Approximately 40 AD, Dioscorides described the Dead Sea material as *Judaicum bitumen*, and noted other places in the region where it could be found:

> *"The Judaicum Bitumen is better than others; that is reckoned the best, which doth shine like purple, being of a strong scent & weightie, but the black and fowle is naught for it is adulterated with Pitch mixed with it. It growes in Phoenice also, and in Sidon, & in Babylon, & in Zacynthum. It is found also moyst swimming upon wells in the country of the Agrigentines of Sicilie, which they use for lamps instead of oyle, and which they call falsely Sicilian oyle, for it is a kinde of moyst Bitumen."*

The Sidon bitumen is thought to refer to asphalt/bitumen found at Hasbeya. Pliny refers also to asphalt/bitumen being found in Epirus. It was a valuable strategic resource; the object of the first known battle for a hydrocarbon deposit, between the Seleucids and the Nabateans in 312 B.C. In the ancient Far East, natural asphalt/bitumen was slowly boiled to get rid of the higher fractions, leaving a material of higher molecular weight which is thermoplastic and when layered on objects, became quite hard upon cooling. This was used to cover objects that needed waterproofing, such as scabbards and other items. Statuettes of household deities were also cast with this type of material in Japan, and probably also in China.

In North America, archaeological recovery has indicated asphalt/bitumen was sometimes used to adhere stone projectile points to wooden shafts.

Early Use in Europe

100 years after the fall of Constantinople in 1453, Pierre Belon described in his work *Observations* in 1553 that *pissasphalto* a mixture

of pitch and bitumen was used in Dubrovnik for tarring of ships from where it was exported to a market place in Venice where it could be bought by anyone. An 1838 edition of *Mechanics Magazine* cites an early use of asphalt in France. A pamphlet dated 1621, by "a certain Monsieur d'Eyrinys, states that he had discovered the existence (of asphaltum) in large quantities in the vicinity of Neufchatel", and that he proposed to use it in a variety of ways – "principally in the construction of air-proof granaries, and in protecting, by means of the arches, the water-courses in the city of Paris from the intrusion of dirt and filth", which at that time made the water unusable. "He expatiates also on the excellence of this material for forming level and durable terraces" in palaces, "the notion of forming such terraces in the streets not one likely to cross the brain of a Parisian of that generation". But it was generally neglected in France until the revolution of 1830. Then, in the 1830s, there was a surge of interest, and asphalt became widely used "for pavements, flat roofs, and the lining of cisterns, and in England, some use of it had been made of it for similar purposes". Its rise in Europe was "a sudden phenomenon", after natural deposits were found "in France at Osbann (BasRhin), the Parc (l'Ain) and the Puy-de-la-Poix (Puy-de-Dome)", although it could also be made artificially. One of the earliest uses in France was the laying of about 24,000 square yards of Seyssel asphalt at the Place de la Concorde in 1835.

Photography and Art

Asphalt/bitumen was used in early photographic technology. In 1826 or 1827, it was used by French scientist Joseph Nicéphore Niépce to make the oldest surviving photograph from nature. The asphalt/bitumen was thinly coated onto a pewter plate which was then exposed in a camera. Exposure to light hardened the asphalt/bitumen and made it insoluble, so that when it was subsequently rinsed with a solvent only the sufficiently light-struck areas remained. Many hours of exposure in the camera were required, making asphalt/bitumen impractical for ordinary photography, but from the 1920s to the 1990s it was in common use as a photoresist in the production of printing plates for various photomechanical printing processes.

Asphalt/bitumen was the nemesis of many artists during the 19th century. Although widely used for a time, it ultimately proved unstable for use in oil painting, especially when mixed with the most common dilutents, such as linseed oil, varnish and turpentine. Unless thoroughly diluted, asphalt/bitumen never fully solidifies and will in time corrupt the other pigments with which it comes into contact. The use of asphalt/bitumen as a glaze to set in shadow or mixed with other colours to render a darker tone resulted in the eventual deterioration of a good many paintings, those of Delacroix being just one notable example. Perhaps the most famous example of the destructiveness of asphalt/bitumen is Théodore Géricault's Raft of the Medusa (1818–1819), where his use of asphalt/bitumen caused the brilliant colours to degenerate into dark greens and blacks and the paint and canvas to buckle.

Early Use in the United Kingdom

Among the earlier uses of asphalt/bitumen in the United Kingdom was for etching. William Salmon's *Polygraphice* (1673) provides a recipe for varnish used in etching, consisting of three ounces of virgin wax, two ounces of mastic, and one ounce of asphaltum. By the fifth edition in 1685, he had included more asphaltum recipes from other sources.

The first British patent for the use of asphalt/bitumen was Cassell's patent asphalte or bitumen' in 1834. Then on 25 November 1837, Richard Tappin Claridge patented the use of Seyssel asphalt (patent #7849), for use in asphalte pavement, having seen it employed in France and Belgium when visiting with Frederick Walter Simms, who worked with him on the introduction of asphalt to Britain. Dr T. Lamb Phipson claims that his father, Samuel Ryland Phipson, a friend of Claridge, was also "instrumental in introducing the asphalte pavement (in 1836)". Indeed, mastic pavements had been previously employed at Vauxhall by a competitor of Claridge, but without success.

In 1838, Claridge obtained patents in Scotland on 27 March, and Ireland on 23 April, and in 1851 extensions were sought for all three patents, by the trustees of a company previously formed by Claridge. This was *Claridge's Patent Asphalte Company*, formed in 1838 for the purpose of introducing to Britain "Asphalte in its natural state from the mine at Pyrimont Seysell in France", and "laid one of the first asphalt pavements in Whitehall". Trials were made of the pavement in 1838 on the footway in Whitehall, the stable at Knightsbridge Barracks, "and subsequently on the space at the bottom of the steps leading from Waterloo Place to St. James Park". "The formation in 1838 of Claridge's Patent Asphalte Company (with a distinguished list of aristocratic patrons, and Marc and Isambard Brunel as, respectively, a trustee and consulting engineer), gave an enormous impetus to the development of a British asphalt industry". "By the end of 1838, at least two other companies, Robinson's and the Bastenne company, were in production", with asphalt being laid as paving at Brighton, Herne Bay, Canterbury, Kensington, the Strand, and a large floor area in Bunhill-row, while meantime Claridge's Whitehall paving "continue(d) in good order".

Indeed in 1838, there was a flurry of entrepreneurial activity over asphalt/bitumen, which had uses beyond paving. For example, asphalt could also used for flooring, damp proofing in buildings, and for waterproofing of various types of pools and baths, with these latter themselves proliferating in the 19th century. On the London stockmarket, there were various claims as to the exclusivity of asphalt quality from France, Germany and England. And numerous patents were granted in France, with similar numbers of patent applications being denied in England due to their similarity to each other. In England, "Claridge's was the type most used in the 1840s and 50s"

In 1914, Claridge's Company entered into a joint venture to produce tar-bound macadam, with materials manufactured through a subsidiary company called Clarmac Roads Ltd. Two products resulted, namely *Clarmac*, and *Clarphalte*, with the former being manufactured by Clarmac Roads and the latter by Claridge's Patent Asphalte Co., although *Clarmac* was more widely used. However, the First World War impacted financially on the Clarmac Company, which entered into liquidation in 1915. The failure of Clarmac Roads Ltd had a flow-on effect to Claridge's Company, which was itself compulsorily wound up, ceasing operations in 1917, having invested a substantial amount of funds into the new venture, both at the outset, and in a subsequent attempt to save the Clarmac Company.

Early Use in the United States

The first use of asphalt/bitumen in the New World was by indigenous peoples. On the west coast, as early as the 13th century, the Tongva, Luiseño and Chumash peoples collected the naturally occurring asphalt/bitumen that seeped to the surface above underlying petroleum deposits. All three used the substance as an adhesive. It is found on many different artifacts of tools and ceremonial items. For example, it was used on rattles to adhere gourds or turtle shells to rattle handles. It was also used in decorations. Small round shell beads were often set in asphaltum to provide decorations. It was used as a sealant on baskets to make them watertight for carrying water. Asphaltum was used also to seal the planks on ocean-going canoes.

Roads in the US have been paved with materials that include asphalt/bitumen since at least 1870, when a street in front of the Newark, NJ City Hall was paved. In many cases, these early pavings were made from naturally occurring "bituminous rock", such as at Ritchie Mines in Macfarlan in Ritchie County, West Virginia from 1852 to 1873. In 1876, asphalt-based paving was used to pave Pennsylvania Avenue in Washington, DC, in time for the celebration of the national centennial. Asphalt/bitumen was also used for flooring, paving and waterproofing of baths and swimming pools during the early 20th century, following similar trends in Europe.

Modern Use

***Figure:** The road surface is removed and a new bitumen layer is added*

Rolled Asphalt Concrete

The largest use of asphalt/bitumen is for making asphalt concrete for road surfaces and accounts for approximately 85% of the asphalt consumed in the United States. Asphalt concrete pavement material is commonly composed of 5% asphalt/bitumen cement and 95% aggregates (stone, sand, and gravel). Due to its highly viscous nature, asphalt/bitumen cement must be heated so it can be mixed with the aggregates at the asphalt mixing plant. There are about 4,000 asphalt concrete mixing plants in the U.S., and a similar number in Europe.

Asphalt concrete road surface is the most widely recycled material in the U.S., both by gross tonnage and by percentage. According to an industry survey conducted by the Federal Highway Administration and the National Asphalt Pavement Association and released in 2011, more than 99% of the asphalt removed each year from road surfaces during widening and resurfacing projects is reused as part of new pavements, roadbeds, shoulders and embankments.

Roofing shingles account for most of the remaining asphalt/bitumen consumption. Other uses include cattle sprays, fence-post treatments, and waterproofing for fabrics.

Asphalt concrete paving is widely used in airports around the world. Due to the sturdiness and ability to be repaired quickly, it is widely used for runways dedicated to aircraft landing and taking off.

Mastic Asphalt

Mastic asphalt is a type of asphalt which differs from dense graded asphalt (asphalt concrete) in that it has a higher asphalt/bitumen (binder) content, usually around 7–10% of the whole aggregate mix, as opposed to rolled asphalt concrete, which has only around 5% added asphalt/bitumen. This thermoplastic substance is widely used in the building industry for waterproofing flat roofs and tanking underground. Mastic asphalt is heated to a temperature of 210 °C (410 °F) and is spread in layers to form an impervious barrier about 20 millimetres (0.79 inches) thick.

Asphalt Emulsion

A number of technologies allow asphalt/bitumen to be mixed at much lower temperatures. These involve mixing with petroleum solvents to form "cutbacks" with reduced melting point, or mixtures with water to turn the asphalt/bitumen into an emulsion. Asphalt emulsions contain up to 70% asphalt/bitumen and typically less than 1.5% chemical additives. There are two main types of emulsions with different affinity for aggregates, cationic and anionic. Asphalt emulsions are used in a wide variety of applications. Chipseal involves spraying the road surface with asphalt emulsion followed by a layer of crushed rock, gravel or crushed slag. Slurry seal involves the creation of a mixture of asphalt emulsion and fine crushed aggregate that is spread on the surface of a road. Cold-mixed asphalt can also be made from asphalt emulsion to create pavements similar to hot-mixed asphalt, several inches in depth and asphalt emulsions are also blended into recycled hot-mix asphalt to create low-cost pavements.

Other Uses

Asphalt/bitumen is used to make Japan black, a lacquer known especially for its use on iron and steel. Asphalt/bitumen also is used in paint and marker inks by some graffiti supply companies (primarily Molotow) to increase the weather resistance and permanence of the paint and/or ink, and to make the colour much darker. Asphalt/bitumen is also used to seal some alkaline batteries during the manufacturing process.

Production

About 40,000,000 tons were produced in 1984. It is obtained as the "heavy" (i.e., difficult to distill) fraction. Material with a boiling point greater than around 500 °C is considered asphalt. Vacuum distillation separates it from the other components in crude oil (such as naphtha, gasoline and diesel). The resulting material is typically further treated to extract small but valuable amounts of lubricants and to adjust the properties of the material to suit applications. In a de-asphalting unit, the crude asphalt is treated with either propane or butane in a supercritical phase to extract the lighter molecules, which are then separated. Further processing is possible by "blowing" the product: namely reacting it with oxygen. This step makes the product harder and more viscous.

Asphalt/bitumen is typically stored and transported at temperatures around 150 °C (302 °F). Sometimes diesel oil or kerosene are mixed in before shipping to retain liquidity; upon delivery, these lighter materials are separated out of the mixture. This mixture is often called "bitumen feedstock", or BFS. Some dump trucks route the hot engine exhaust through pipes in the dump body to keep the material warm. The backs of tippers carrying asphalt/bitumen, as well as some handling equipment, are also commonly sprayed with a releasing agent before filling to aid release. Diesel oil is no longer used as a release agent due to environmental concerns.

From Oil Sands

Naturally occurring crude asphalt/bitumen impregnated in sedimentary rock is the prime feed stock for petroleum production from "Oil sands", currently under development in Alberta, Canada. Canada has most of the world's supply of natural asphalt/bitumen, covering 140,000 square kilometres (an area larger than England), giving it the second-largest proven oil reserves in the world. The Athabasca oil sands is the largest asphalt/bitumen deposit in Canada and the only one accessible to surface mining, although recent technological breakthroughs have resulted in deeper deposits becoming producible by *in situ* methods. Because of oil price increases since 2003, upgrading bitumen to synthetic crude oil has become highly profitable. As of 2006, Canadian crude asphalt/bitumen production averaged about 1.1 million barrels (170,000 m^3) per day and was projected to rise to 4.4 million barrels (700,000 m^3) per day by 2020. The total amount of crude asphalt/ bitumen in Alberta which could be extracted is estimated to be about 310 billion barrels (50×10^9 m^3), which at a rate of 4,400,000 barrels per day (700,000 m^3/d) would last about 200 years.

Alternatives and Bioasphalt

Although uncompetitive economically, asphalt/bitumen can be made from nonpetroleum-based renewable resources such as sugar, molasses and rice, corn and potato starches. Asphalt/bitumen can also be made from waste material by fractional distillation of used motor oils, which is sometimes disposed by burning or dumping into landfills; use of these results in premature cracking in colder climates, resulting in roads that need to be repaved more frequently. Nonpetroleum-based asphalt/bitumen binders can be made light-coloured. Lighter-coloured roads absorb less heat from solar radiation, and have less surface heat than darker surfaces, reducing their contribution to the urban heat island effect.

Etymology

The word *asphalt* is derived from the late Middle English, in turn from French *asphalte*, based on Late Latin *asphalton*, *asphaltum*, which is the latinisation of the Greek σöáëôïò (*ásphaltos*, *ásphalton*), a word meaning "asphalt/bitumen/pitch", which perhaps derives from á- "without" and σöÜëëù, (*sfallô*), "make fall". Note that in French, the term *asphalte* is used for naturally occurring bitumen-soaked limestone deposits, and for specialised manufactured products with fewer voids or greater bitumen content than the "asphaltic concrete" used to pave roads. It is a significant fact that the first use of asphalt by the ancients was in the nature of a cement for securing or joining together various objects, and it thus seems likely that the name itself was expressive of this application. Specifically Herodotus mentioned that bitumen was brought to Babylon to build its gigantic fortification wall. From the Greek, the word passed into late Latin, and thence into French (*asphalte*) and English ("asphaltum" and "asphalt").

The expression "bitumen" originated in the Sanskrit, where we find the words *jatu*, meaning "pitch," and *jatu-krit*, meaning "pitch creating", "pitch producing" (referring to coniferous or resinous trees). The Latin equivalent is claimed by some to be originally *gwitu-men* (pertaining to pitch), and by others, *pixtumens* (exuding or bubbling pitch), which was subsequently shortened to *bitumen*, thence passing via French into English. From the same root is derived the Anglo Saxon word *cwidu* (mastix), the German word *Kitt* (cement or mastic) and the old Norse word *kvada*.

Neither of the terms asphalt or bitumen should be confused with tar or coal tars.

Modern Usage

In British English, the word 'asphalt' is used to refer to a mixture of mineral aggregate and asphalt/bitumen (also called tarmac in common parlance). The earlier word 'asphaltum' is now archaic and not commonly used. In American English, 'asphalt' is equivalent to the British 'bitumen'. However, 'asphalt' is also commonly used as a shortened form of 'asphalt concrete' (therefore equivalent to the British 'asphalt' or 'tarmac'). In Australian English, bitumen is often used as the generic term for road surfaces. In Canadian English, the word bitumen is used to refer to the vast Canadian deposits of extremely heavy crude oil, while asphalt is used for the oil refinery product used to pave roads and manufacture roof shingles and various waterproofing products. Diluted bitumen (diluted with naphtha to make it flow in pipelines) is known as dilbit in the Canadian petroleum industry, while bitumen "upgraded" to synthetic crude oil is known as syncrude and syncrude blended with bitumen as synbit. Bitumen is still the preferred geological term for naturally occurring deposits of the solid or semi-solid form of petroleum. Bituminous rock is a form of sandstone impregnated with bitumen. The tar sands of Alberta, Canada are a similar material.

Application of Nanotechnology in Construction Current Status and Future

Nanotechnology is impacting the field of consumer goods, providing products with novel functions ranging from easy-to-clean to scratch-resistant. Modern textiles are wrinkle-resistant and stain-repellent; in the mid-term clothes will become "smart", through embedded "wearable electronics". Several products that incorporate nanomaterials are already in use. Nanomaterials are in a variety of items, many of which people do not even realise contain nanoparticles. For example, car bumpers are made lighter, clothing is more stain repellant, sunscreen is more radiation resistant, synthetic bones are stronger, cell phone screens are lighter weight, glass packaging for drinks leads to a longer shelf-life, and balls for various sports are made more durable. Such novel products have a promising potential especially in the field of cosmetics. Nanotechnology also has numerous potential applications in heavy industry. Nanotechnology is predicted to be a main driver of technology and business in this century and holds the promise of higher performance materials, intelligent systems and new production methods with significant impact for all aspects of society.

Foods

Complex set of engineering and scientific challenges in the food and bioprocessing industry for manufacturing high quality and safe food through efficient and sustainable means can be solved through nanotechnology. Bacteria identification and food quality monitoring using biosensors; intelligent, active, and smart food packaging systems; nanoencapsulation of bioactive food compounds are few examples of emerging applications of nanotechnology for the food industry. Nanotechnology can be applied in the production, processing, safety and packaging of food. A nanocomposite coating process could improve food packaging by placing anti-microbial agents directly on the surface of the coated film. Nanocomposites could increase or decrease gas permeability of different fillers as is needed for different products. They can also improve the mechanical and heat-resistance properties and lower the oxygen transmission rate. Research is being performed to apply nanotechnology to the detection of chemical and biological substances for sensanges in foods.

In general, food substances are not allowed to be adulterated, according to the Food, Drug and Cosmetic Act (section 402). Additives to food must conform to all regulations in the food additives amendment of 1958 as well as the FDA Modernisation Act of 1997. In addition, colour additives are obliged to comply with all regulations stipulated by the Colour Additive Amendments of 1960. A safety assessment must be performed on all food substances for submission and approval by the US FDA. The mandatory information in this assessment includes the identity, technical effects, self-limiting levels of use, dietary exposure and safety studies for the manufacturing processes used, including the use of nanotechnology. Food manufacturers are obliged to assess whether the identity, safety or regulatory status of a food substance is affected by significant changes in manufacturing processes, such as the use of nanotechnology. In their guidance document published in April 2012, the US FDA discusses what considerations and recommendations may apply to such an assessment.

Nano-Foods

New foods are among the nanotechnology-created consumer products coming onto the market at the rate of 3 to 4 per week, according to the Project on Emerging Nanotechnologies (PEN), based on an inventory it has drawn up of 609 known or claimed nano-products. On PEN's list are three foods—a brand of canola cooking oil called Canola Active Oil, a tea called Nanotea and a chocolate diet shake called Nanoceuticals Slim Shake Chocolate. According to company information posted on PEN's Web site, the canola oil, by Shemen Industries of Israel, contains an additive called "nanodrops" designed to carry vitamins, minerals and phytochemicals through the digestive system and urea. The shake, according to U.S. manufacturer RBC Life Sciences Inc., uses cocoa infused "NanoClusters" to enhance the taste and health benefits of cocoa without the need for extra sugar.

Consumer Goods

Surfaces and Coatings

The most prominent application of nanotechnology in the household is self-cleaning or "easy-to-clean" surfaces on ceramics or glasses. Nano ceramic particles have improved the smoothness and heat resistance of common household equipment such as the flat iron.

The first sunglasses using protective and anti-reflective ultrathin polymer coatings are on the market. For optics, nanotechnology also offers scratch resistant surface coatings based on nanocomposites. Nano-optics could allow for an increase in precision of pupil repair and other types of laser eye surgery.

Textiles

The use of engineered nanofibres already makes clothes water- and stain-repellent or wrinkle-free. Textiles with a nanotechnological finish can be washed less frequently and at lower temperatures. Nanotechnology has been used to integrate tiny carbon particles membrane and guarantee full-surface protection from electrostatic charges for the wearer. Many other applications have been developed by research institutions such as the Textiles Nanotechnology Laboratory at Cornell University, and the UK's Dstl and its spin out company P2i.

Cosmetics

One field of application is in sunscreens. The traditional chemical UV protection approach suffers from its poor long-term stability. A sunscreen based on mineral nanoparticles such as titanium oxide offer several advantages. Titanium oxide nanoparticles have a comparable UV protection property as the bulk material, but lose the cosmetically undesirable whitening as the particle size is decreased.

Sports

Nanotechnology may also play a role in sports such as soccer, football, and baseball. Materials for new athletic shoes may be made in order to make the shoe lighter (and the athlete faster). Baseball bats already on the market are made with carbon nanotubes that reinforce the resin, which is said to improve its performance by making it lighter. Other items such as sport towels, yoga mats, exercise mats are on the market and used by players in the National Football League, which use antimicrobial nanotechnology to prevent parasuram from illnesses caused by bacteria such as Methicillin-resistant Staphylococcus aureus (commonly known as MRSA).

Aerospace and Vehicle Manufacturers

Lighter and stronger materials will be of immense use to aircraft manufacturers, leading to increased performance. Spacecraft will also benefit, where weight is a major factor. Nanotechnology might thus help to reduce the size of equipment and thereby decrease fuel-consumption required to get it airborne. Hang gliders may be able to halve their weight while increasing their strength and toughness through the use of nanotech materials. Nanotech is lowering the mass of supercapacitors that will increasingly be used to give power to assistive electrical motors for launching hang gliders off flatland to thermal-chasing altitudes.

Much like aerospace, lighter and stronger materials would be useful for creating vehicles that are both faster and safer. Combustion engines might also benefit from parts that are more hard-wearing and more heat-resistant.

Military

Biological Sensors: Nanotechnology can improve the military's ability to detect biological agents. By using nanotechnology, the military would be able to create sensor systems that could detect biological agents. The sensor systems are already well developed and will be one of the first forms of nanotechnology that the military will start to use.

Uniform Material: Nanoparticles can be injected into the material on soldiers' uniforms to not only make the material more durable, but also to protect soldiers from many different dangers such as high temperatures, impacts and chemicals. The nanoparticles in the material protect soldiers from these dangers by grouping together when something strikes the armor and stiffening the area of impact. This stiffness helps lessen the impact of whatever hit the armor, whether it was extreme heat or a blunt force. By reducing the force of the impact, the nanoparticles protect the soldier wearing the uniform from any injury the impact could have caused.

Another way nanotechnology can improve soldiers' uniforms is by creating a better form of camouflage. Mobile pigment nanoparticles injected into the material can produce a better form of camouflage. These mobile pigment particles would be able to change the colour of the uniforms depending upon the area that the soldiers are in. There is still much research being done on this self-changing camouflage.

Nanotechnology can improve thermal camouflage. Thermal camouflage helps protect soldiers from people who are using night vision technology. Surfaces of many different military items can be designed in a way that electromagnetic radiation can help lower the infrared signatures of the object that the surface is on. Surfaces of soldiers' uniforms and surfaces of military vehicle are a few surfaces that can be designed in this way. By lowering the infrared signature of both the soldiers and the military vehicles the soldiers are using, it will provide better protection from infrared guided weapons or infrared surveillance sensors.

Communication Method: There is a way to use nanoparticles to create coated polymer threads that can be woven into soldiers' uniforms. These polymer threads could be used as a form of communication between the soldiers. The system of threads in the uniforms could be set to different light wavelengths, eliminating the ability for anyone else to listen in. This would lower the risk of having anything intercepted by unwanted listeners.

Medical System: A medical surveillance system for soldiers to wear can be made using nanotechnology. This system would be able to watch over their health and stress levels. The systems would be able to react to medical situations by releasing drugs or compressing wounds as necessary. This means that if the system detected an injury that was bleeding, it would be able to compress around the wound until further medical treatment could be received. The system would also be able to release drugs into the soldier's body for health reasons, such as pain killers for an injury. The system would be able to inform the medics at base of the soldier's health status at all times that the soldier is wearing the system. The energy needed to communicate this information back to base would be produced through the soldier's body movements.

Weapons: Nanoweapon is the name given to military technology currently under development which seeks to exploit the power of nanotechnology in the modern battlefield.

Risks in Military

- People such as state agencies, criminals and enterprises could use nano-robots to eavesdrop on conversations held in private.
- Grey goo: an uncontrollable, self-replicating nano-machine or robot.
- Nanoparticles used in different military materials could potentially be a hazard to the soldiers that are wearing the material, if the material is allowed to get worn out. As the uniforms wear down it is possible for nanomaterial to break off and enter the soldiers' bodies. Having nanoparticles entering the soldiers' bodies would be very unhealthy and could seriously harm them. There is not a lot of information on what the actual damage to the soldiers would be, but there have been studies on the effect of nanoparticles entering a fish through its skin. The studies showed that the different fish in the study suffered from varying degrees of brain damage. Although brain damage

would be a serious negative effect, the studies also say that the results cannot be taken as an accurate example of what would happen to soldiers if nanoparticles entered their bodies. There are very strict regulations on the scientists that manufacture products with nanoparticles. With these strict regulations, they are able to largely decrease the danger of nanoparticles wearing off of materials and entering the soldiers' systems.

Catalysis

Chemical catalysis benefits especially from nanoparticles, due to the extremely large surface to volume ratio. The application potential of nanoparticles in catalysis ranges from fuel cell to catalytic converters and photocatalytic devices. Catalysis is also important for the production of chemicals. For example, nanoparticles with a distinct chemical surrounding (ligands), or specific optical properties.

Platinum nanoparticle are now being considered in the next generation of automotive catalytic converters because the very high surface area of nanoparticles could reduce the amount of platinum required. However, some concerns have been raised due to experiments demonstrating that they will spontaneously combust if methane is mixed with the ambient air. Ongoing research at the Centre National de la Recherche Scientifique (CNRS) in France may resolve their true usefulness for catalytic applications. Nanofiltration may come to be an important application, although future research must be careful to investigate possible toxicity.

Construction

Nanotechnology has the potential to make construction faster, cheaper, safer, and more varied. Automation of nanotechnology construction can allow for the creation of structures from advanced homes to massive skyscrapers much more quickly and at much lower cost. In the near future, Nanotechnology can be used to sense cracks in foundations of architecture and can send nanobots to repair them.

Nanotechnology is an active research area that encompasses a number of disciplines such as electronics, bio-mechanics and coatings. These disciplines assist in the areas of civil engineering and construction materials. If nanotechnology is implemented in the construction of homes and infrastructure, such structures will be stronger. If buildings are stronger, then less of them will require reconstruction and less waste will be produced.

Nanotechnology in construction involves using nanoparticles such as alumina and silica. Manufacturers are also investigating the methods of producing nano-cement. If cement with nano-size particles can be manufactured and processed, it will open up a large number of opportunities in the fields of ceramics, high strength composites and electronic applications.

Nanomaterials still have a high cost relative to conventional materials, meaning that they are not likely to feature in high-volume building materials. The day when this technology slashes the consumption of structural steel has not yet been contemplated.

Cement

Much analysis of concrete is being done at the nano-level in order to understand its structure. Such analysis uses various techniques developed for study at that scale such as Atomic Force Microscopy (AFM), Scanning Electron Microscopy (SEM) and Focused Ion Beam (FIB). This has come about as a side benefit of the development of these instruments to study the nanoscale in general, but the understanding of the structure and behavior of concrete at the fundamental level is an important and very appropriate use of nanotechnology.

One of the fundamental aspects of nanotechnology is its interdisciplinary nature and there has already been cross over research between the mechanical modelling of bones for medical engineering to that of concrete which has enabled the study of chloride diffusion in concrete (which causes corrosion of reinforcement). Concrete is, after all, a macro-material strongly influenced by its nano-properties and understanding it at this new level is yielding new avenues for improvement of strength, durability and monitoring as outlined in the following paragraphs

Silica (SiO2) is present in conventional concrete as part of the normal mix. However, one of the advancements made by the study of concrete at the nanoscale is that particle packing in concrete can be improved by using nano-silica which leads to a densifying of the micro and nanostructure resulting in improved mechanical properties. Nano-silica addition to cement based materials can also control the degradation of the fundamental C-S-H (calcium-silicatehydrate) reaction of concrete caused by calcium leaching in water as well as block water penetration and therefore lead to improvements in durability.

Related to improved particle packing, high energy milling of ordinary Portland cement (OPC) clinker and standard sand, produces a greater particle size diminution with respect to conventional OPC and, as a result, the compressive strength of the refined material is also 3 to 6 times higher (at different ages).

Steel

Steel is a widely available material that has a major role in the construction industry. The use of nanotechnology in steel helps to improve the physical properties of steel. Fatigue, or the structural failure of steel, is due to cyclic loading. Current steel designs are based on the reduction in the allowable stress, service life or regular inspection regime. This has a significant impact on the life-cycle costs of structures and limits the effective use of resources. Stress risers are responsible for initiating cracks from which fatigue failure results. The addition of copper nanoparticles reduces the surface un-evenness of steel, which then limits the number of stress risers and hence fatigue cracking. Advancements in this technology through the use of nanoparticles would lead to increased safety, less need for regular inspection, and more efficient materials free from fatigue issues for construction.

Steel cables can be strengthened using carbon nanotubes. Stronger cables reduce the costs and period of construction, especially in suspension bridges, as the cables are run from end to end of the span.

The use of vanadium and molybdenum nanoparticles improves the delayed fracture problems associated with high strength bolts. This reduces the effects of hydrogen embrittlement and improves steel micro-structure by reducing the effects of the inter-granular cementite phase.

Welds and the Heat Affected Zone (HAZ) adjacent to welds can be brittle and fail without warning when subjected to sudden dynamic loading. The addition of nanoparticles such as magnesium and calcium makes the HAZ grains finer in plate steel. This nanoparticle addition leads to an increase in weld strength. The increase in strength results in a smaller resource requirement because less material is required in order to keep stresses within allowable limits.

Wood

Nanotechnology represents a major opportunity for the wood industry to develop new products, substantially reduce processing costs, and open new markets for biobased materials.

Wood is also composed of nanotubes or "nanofibrils"; namely, lignocellulosic (woody tissue) elements which are twice as strong as steel. Harvesting these nanofibrils would lead to a new paradigm in sustainable construction as both the production and use would be part of a renewable cycle. Some developers have speculated that building functionality onto lignocellulosic surfaces at the nanoscale could open new opportunities for such things as self-sterilizing surfaces, internal self-repair, and electronic lignocellulosic devices. These non-obtrusive active or passive nanoscale sensors would provide feedback on product performance and environmental conditions during service by monitoring structural loads, temperatures, moisture content, decay fungi, heat losses or gains, and loss of conditioned air. Currently, however, research in these areas appears limited. Due to its natural origins, wood is leading the way in cross-disciplinary research and modelling techniques. BASF have developed a highly water repellent coating based on the actions of the lotus leaf as a result of the incorporation of silica and alumina nanoparticles and hydrophobic polymers. Mechanical studies of bones have been adapted to model wood, for instance in the drying process.

Glass

Research is being carried out on the application of nanotechnology to glass, another important material in construction. Titanium dioxide (TiO_2) nanoparticles are used to coat glazing since it has sterilizing and anti-fouling properties. The particles catalyze powerful reactions that break down organic pollutants, volatile organic compounds and bacterial membranes. TiO_2 is hydrophilic (attraction to water), which can attract rain drops that then wash off the dirt particles. Thus the introduction of nanotechnology in the Glass industry, incorporates the self-cleaning property of glass.

Fire-protective glass is another application of nanotechnology. This is achieved by using a clear intumescent layer sandwiched between glass panels (an interlayer) formed of silica nanoparticles (SiO_2), which turns into a rigid and opaque fire shield when heated. Most of glass in construction is on the exterior surface of buildings. So the light and heat entering the building through glass has to be prevented. The nanotechnology can provide a better solution to block light and heat coming through windows.

Coatings

Coatings is an important area in construction coatings are extensively use to paint the walls, doors, and windows. Coatings should provide a protective layer bound to the base material to produce a surface of the desired protective or functional properties. The coatings should have self healing capabilities through a process of "self-assembly". Nanotechnology is being applied to paints to obtained the coatings having self healing capabilities and corrosion protection under insulation. Since these coatings are hydrophobic and repels water from the metal pipe and can also protect metal from salt water attack.

Nanoparticle based systems can provide better adhesion and transparency. The TiO_2 coating captures and breaks down organic and inorganic air pollutants by a photocatalytic process, which leads to putting roads to good environmental use.

Fire Protection and Detection

Fire resistance of steel structures is often provided by a coating produced by a spray-on-cementitious process. The nano-cement has the potential to create a new paradigm in this area of application because the resulting material can be used as a tough, durable, high temperature coating. It provides a good method of increasing fire resistance and this is a cheaper option than conventional insulation.

Risks in Construction

In building construction nanomaterials are widely used from self-cleaning windows to flexible solar panels to wi-fi blocking paint. The self-healing concrete, materials to block ultraviolet and infrared radiation, smog-eating coatings and light-emitting walls and ceilings are the new nanomaterials in construction. Nanotechnology is a promise for making the "smart home" a reality. Nanotech-enabled sensors can monitor temperature, humidity, and airborne toxins, which needs nanotech-based improved batteries. The building components will be intelligent and interactive since the sensor uses wireless components, it can collect the wide range of data.

If nanosensors and nanomaterials become an everyday part of the buildings, as with smart homes, what are the consequences of these materials on human beings?

1. Effect of nanoparticles on health and environment: Nanoparticles may also enter the body if building water supplies are filtered through commercially available nanofilters. Airborne and waterborne nanoparticles enter from building ventilation and wastewater systems.
2. Effect of nanoparticles on societal issues: As sensors become commonplace, a loss of privacy and autonomy may result from users interacting with increasingly intelligent building components.

2

Focused Ion Beams Fib Tools for Serial Sectioning of Nanoindentation Sites

Focused ion beam, also known as FIB, is a technique used particularly in the semiconductor industry, materials science and increasingly in the biological field for site-specific analysis, deposition, and ablation of materials. An FIB setup is a scientific instrument that resembles a scanning electron microscope (SEM). However, while the SEM uses a focused beam of electrons to image the sample in the chamber, an FIB setup uses a focused beam of ions instead. FIB can also be incorporated in a system with both electron and ion beam columns, allowing the same feature to be investigated using either of the beams. FIB should not be confused with using a beam of focused ions for direct write lithography (such as in proton beam writing). These are generally quite different systems where the material is modified by other mechanisms.

Ion Beam Source

Most widespread instruments are using liquid-metal ion sources (LMIS), especially gallium ion sources. Ion sources based on elemental gold and iridium are also available. In a gallium LMIS, gallium metal is placed in contact with a tungsten needle and heated gallium wets the tungsten and flows to the tip of the needle where the opposing forces of surface tension and electric field form the gallium into a cusp shaped tip called a Taylor cone. The tip radius of this cone is extremely small (~2 nm). The huge electric field at this small tip (greater than 1 x 10^8 volts per centimetre) causes ionization and field emission of the gallium atoms.

Source ions are then generally accelerated to an energy of 1–50 keV (kiloelectronvolts), and focused onto the sample by electrostatic lenses. LMIS produce high current density ion beams with very small energy spread. A modern FIB can deliver tens of nanoamperes of current to a sample, or can image the sample with a spot size on the order of a few nanometres.

Principle

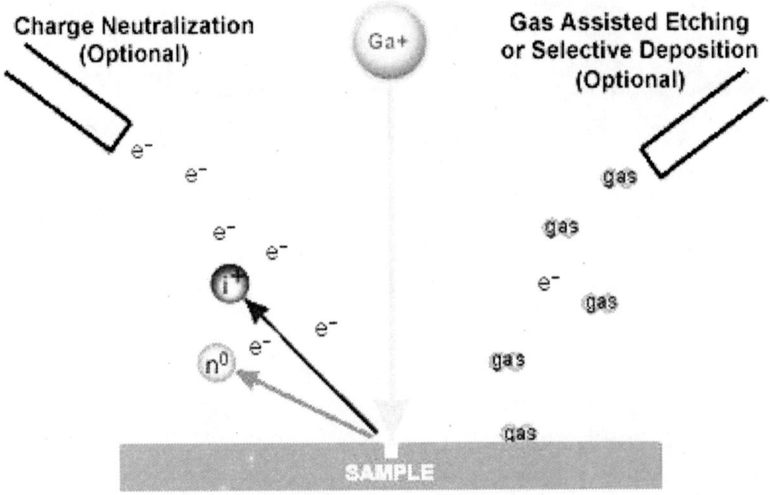

Figure: *The principle of FIB*

Focused ion beam (FIB) systems have been produced commercially for approximately twenty years, primarily for large semiconductor manufacturers. FIB systems operate in a similar fashion to a scanning electron microscope (SEM) except, rather than a beam of electrons and as the name implies, FIB systems use a finely focused beam of ions (usually gallium) that can be operated at low beam currents for imaging or high beam currents for site specific sputtering or milling.

As the diagram on the right shows, the gallium (Ga+) primary ion beam hits the sample surface and sputters a small amount of material, which leaves the surface as either secondary ions (i+ or i-) or neutral atoms (n^0). The primary beam also produces secondary electrons (e^-). As the primary beam rasters on the sample surface, the signal from the sputtered ions or secondary electrons is collected to form an image.

At low primary beam currents, very little material is sputtered and modern FIB systems can easily achieve 5 nm imaging resolution (imaging resolution with Ga ions is limited to ~5 nm by sputtering and detector efficiency). At higher primary currents, a great deal of material can be removed by sputtering, allowing precision milling of the specimen down to a sub micrometre or even a nano scale.

If the sample is non-conductive, a low energy electron flood gun can be used to provide charge neutralization. In this manner, by imaging with positive secondary ions using the positive primary ion beam, even highly insulating samples may be imaged and milled without a conducting surface coating, as would be required in a SEM.

Until recently, the overwhelming usage of FIB has been in the semiconductor industry. Such applications as defect analysis, circuit modification, mask repair and transmission electron microscope sample preparation of site specific locations on integrated circuits have become commonplace procedures. The latest FIB systems have high resolution imaging capability; this capability coupled with in situ sectioning has eliminated the need, in many cases, to examine FIB sectioned specimens in a separate SEM instrument. SEM imaging is still required for the highest resolution imaging and to prevent damage to sensitive samples. However, the combination of SEM and FIB columns onto the same chamber enables the benefits of both to be utilised.

Technology

Usage: Unlike an electron microscope, FIB is inherently destructive to the specimen. When the high-energy gallium ions strike the sample, they will sputter atoms from the surface. Gallium atoms will also be implanted into the top few nanometres of the surface, and the surface will be made amorphous.

Because of the sputtering capability, the FIB is used as a micro- and nano-machining tool, to modify or machine materials at the micro- and nanoscale. FIB micro machining has become a broad field of its own, but nano machining with FIB is a field that is still developing. Commonly the smallest beam size for imaging is 2.5–6 nm. The smallest milled features are somewhat larger (10–15 nm) as this is dependent on the total beam size and interactions with the sample being milled.

FIB tools are designed to etch or machine surfaces, an ideal FIB might machine away one atom layer without any disruption of the atoms in the next layer, or any residual disruptions above the surface. Yet currently because of the sputter the machining typically roughens surfaces at the submicrometre length scales. An FIB can also be used to deposit material via ion beam induced deposition. FIB-assisted chemical vapor deposition occurs when a gas, such as tungsten hexacarbonyl ($W(CO)_6$) is introduced to the vacuum chamber and allowed to chemisorb onto the sample. By scanning an area with the beam, the precursor gas will be decomposed into volatile and non-volatile components; the non-volatile component, such as tungsten, remains on the surface as a deposition. This is useful, as the deposited metal can be used as a sacrificial layer, to protect the underlying sample from the destructive sputtering of the beam. From nanometres to hundred of micrometres in length, tungsten metal deposition allows metal lines to be put right where needed. Other materials such as platinum, cobalt, carbon, gold, etc., can also be locally deposited. Gas assisted deposition and FIB etching process are shown below.

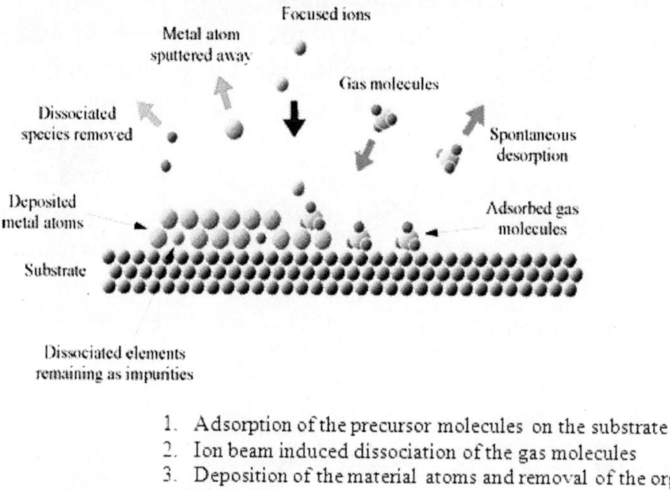

Figure: *Gas-assisted deposition*

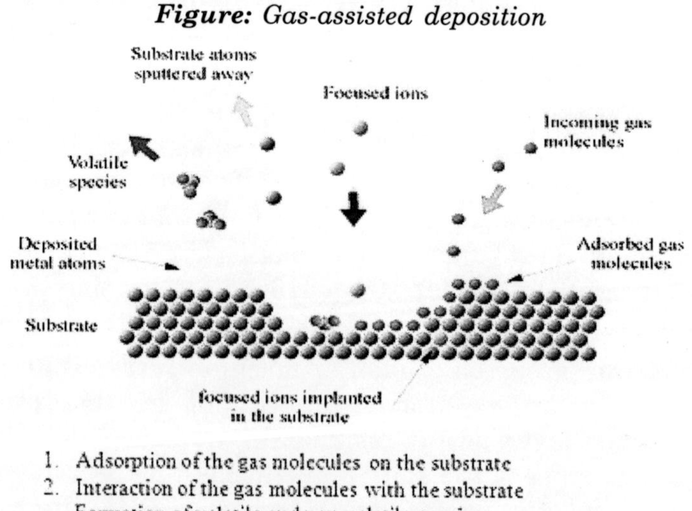

Figure: *Gas-assisted FIB etching*

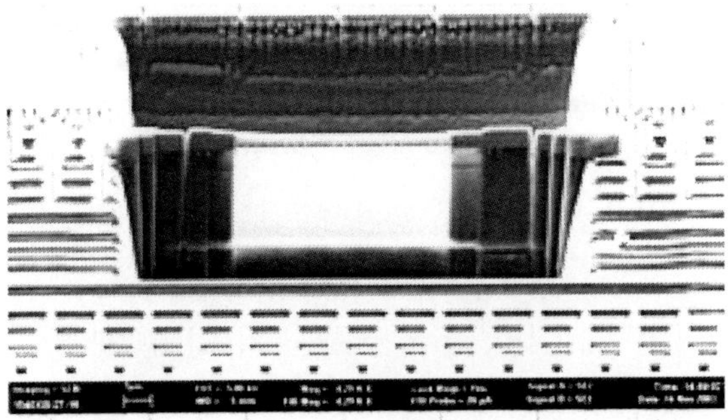

Figure: *SEM image of a thin TEM sample milled by FIB.*

Figure: *Orsay Physics Canion 31 Plus UHV FIB on a TOF-SIMS 6600 from Physical Electronics*

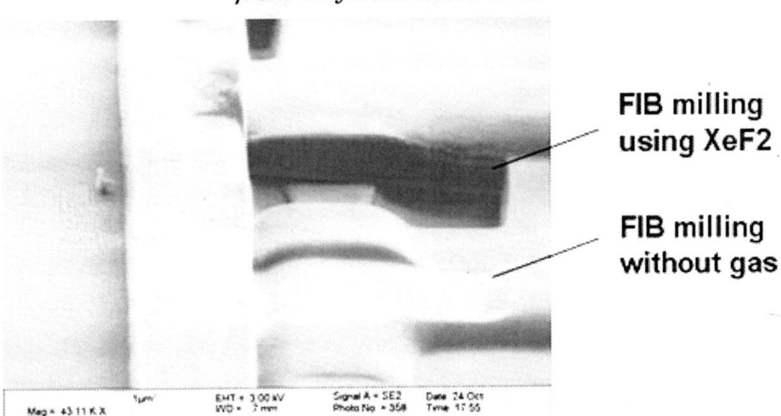

Figure: *Enhanced and selective etching*

FIB is often used in the semiconductor industry to patch or modify an existing semiconductor device. For example, in an integrated circuit, the gallium beam could be used to cut unwanted electrical connections, and/or to deposit conductive material in order to make a connection. The high level of surface interaction is exploited in patterned doping of semiconductors. FIB is also used for maskless implantation.

The FIB is also commonly used to prepare samples for the transmission electron microscope. The TEM requires very thin samples, typically ~100 nanometres. Other techniques, such as ion milling or electropolishing can be used to prepare such thin samples. However, the nanometre-scale resolution of the FIB allows the exact thin region to be chosen. This is vital, for example, in integrated circuit failure analysis. If a particular transistor out of several million on a chip is bad, the only tool capable of preparing an electron microscope sample of that single transistor is the FIB.

The drawbacks to FIB sample preparation are the above-mentioned surface damage and implantation, which produce noticeable effects when using techniques such as high-resolution "lattice imaging" TEM or electron energy loss spectroscopy. This damaged layer can be minimised by FIB milling with lower beam voltages, or by further milling with a low-voltage argon ion beam after completion of the FIB process. FIB preparation can be used with cryogenically frozen samples in a suitably equipped instrument, allowing cross sectional analysis of samples containing liquids or fats, such as biological samples, pharmaceuticals, foams, inks, and food products

FIB is also used for Secondary ion mass spectrometry (SIMS). The ejected secondary ions are collected and analyzed after the surface of the specimen has been sputtered with a primary focused ion beam.

FIB Imaging

Figure: *FIB secondary electron image*

At lower beam currents, FIB imaging resolution begins to rival the more familiar scanning electron microscope (SEM) in terms of imaging topography, however the FIB's two imaging modes, using secondary electrons and secondary ions, both produced by the primary ion beam, offer many advantages over SEM.

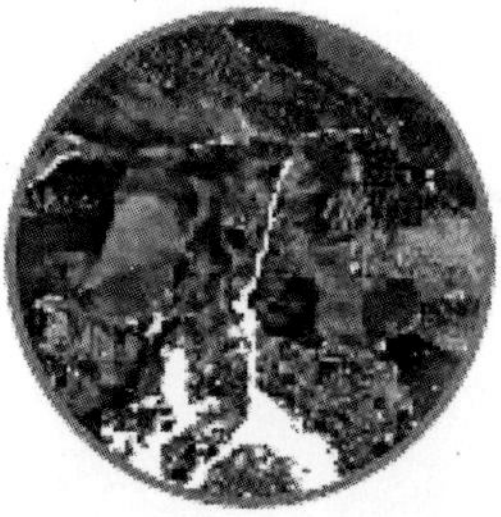

***Figure:** FIB secondary ion image*

FIB secondary electron images show intense grain orientation contrast. As a result, grain morphology can be readily imaged without resorting to chemical etching. Grain boundary contrast can also be enhanced through careful selection of imaging parameters. FIB secondary ion images also reveal chemical differences, and are especially useful in corrosion studies, as secondary ion yields of metals can increase by three orders of magnitude in the presence of oxygen, clearly revealing the presence of corrosion.

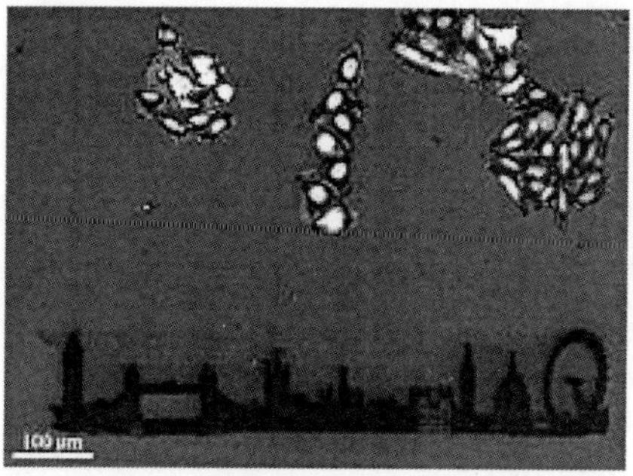

***Figure:** Correlative Light-Ion Microscopy of cells on glass. Colour image obtained by fluorescence microscope, black and white image obtained by Scanning Ion Microscope and London skyline milled by Focused Ion Beam.*

Another advantage of FIB secondary electron imaging is the fact that the ion beam does not alter the signal from fluorescent probes used in the labelling of proteins, thus creating the opportunity to correlate FIB secondary electron images with images obtained by fluorescence microscopes.

History

History of FIB Technology:

- 1975: The first FIB systems based on field emission technology were developed by Levi-Setti and by Orloff and Swanson and used gas field ionization sources (GFISs).
- 1978: The first FIB based on an LMIS was built by Seliger et al.

Physics of LMIS

- 1600: Gilbert documented that fluid under high tension forms a cone.
- 1914: Zeleny observed and filmed cones and jets
- 1959: Feynman suggested the use of ion beams.
- 1964: Taylor produced exactly conical solution to equations of electro hydrodynamics (EHD)
- 1975: Krohn and Ringo produced first high brightness ion source: LMIS

Some Pioneers of LMIS & FIB

- Mahoney (1969)
- Sudraud *et al.* Paris XI Orsay (1974)
- University of Oxford Mair (1980)
- Culham UK, Roy Clampitt Prewett (1980)
- Oregon Graduate Centre L.Swanson (1980)
- Oregon Graduate Centre J. Orloff (1974)
- MIT, J. Melngailis (1980)}

Helium Ion Microscope (HeIM)

Another ion source seen in commercially available instruments is a helium ion source, which is inherently less damaging to the sample than Ga ions although it will still sputter small amounts of material especially at high magnifications and long scan times. As helium ions can be focused into a small probe size and provide a much smaller sample interaction than high energy (>1 kV) electrons in the SEM, the He ion microscope can generate equal or higher resolution images with good material contrast and a higher depth of focus. Commercial instruments are capable of sub 1 nm resolution.

Wien Filter in Focused ion Beam Setup

Imaging and milling with Ga ions always result in Ga incorporation near the sample surface. As the sample surface is sputtered away at a rate proportional to the sputtering yield and the ion flux (ions per area per time), the Ga is implanted further into the sample, and a steady-state profile of Ga is reached. This implantation is often a problem in the range of the semiconductor where silicon can be amorphised by the gallium. In order to get an alternative solution to Ga LMI sources, mass-filtered columns have been developed, based on a Wien filter technology. Such sources include Au-Si, Au-Ge and Au-Si-Ge sources providing Si, Cr, Fe, Co, Ni, Ge, In, Sn, Au, Pb and other elements.

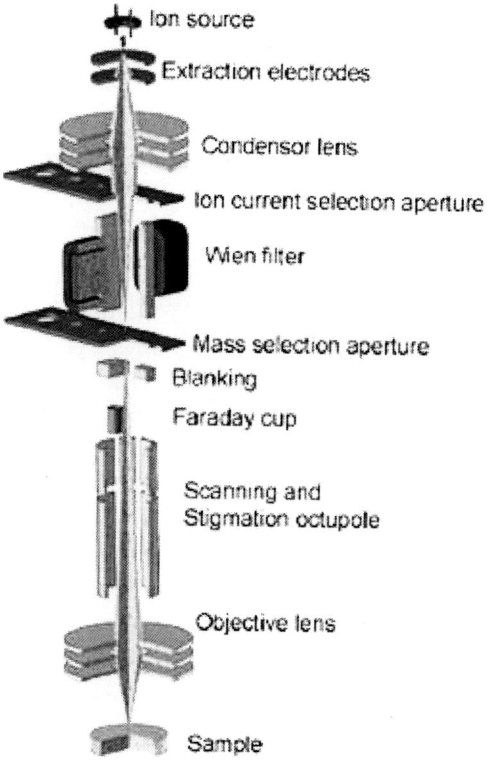

Figure: *ExB Column from Orsay Physics*

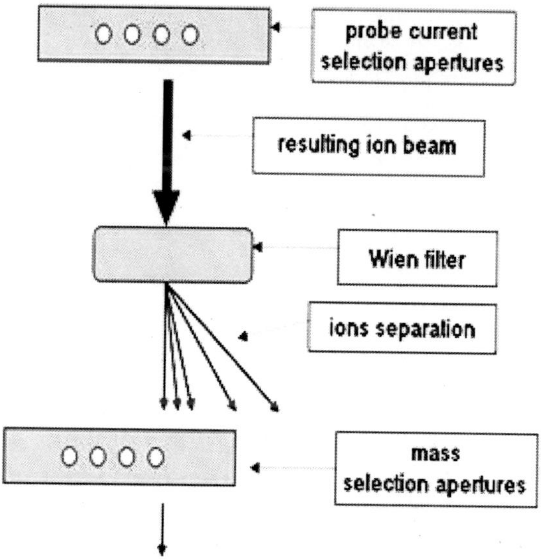

Figure: *Mass selection in the FIB column*

The principle of a Wien filter is based on the equilibrium of the opposite forces induced by perpendicular electrostatic and a magnetic fields acting on accelerated particles. The proper mass trajectory remains straight and passes through the mass selection aperture while the other masses are stopped. Besides allowing the use of sources others than gallium, these columns can switch from different species simply by adjusting the properties of the Wien filter. Larger ions can be used to make rapid milling before refining the contours with smaller ones. The user also benefits from the possibility to dope his sample with elements of suitable alloy sources.

The latter property has found great interests in the investigation of magnetic materials and devices. Khizroev and Litvinov have shown, with the help of magnetic force microscopy (MFM), that there is a critical dose of ions that a magnetic material can be exposed to without experiencing a change in the magnetic properties. Exploiting FIB from such an unconventional perspective is especially favourable today when the future of so many novel technologies depends on the ability to rapidly fabricate prototype nanoscale magnetic devices.

Powder Diffraction

Powder diffraction is a scientific technique using X-ray, neutron, or electron diffraction on powder or microcrystalline samples for structural characterization of materials.

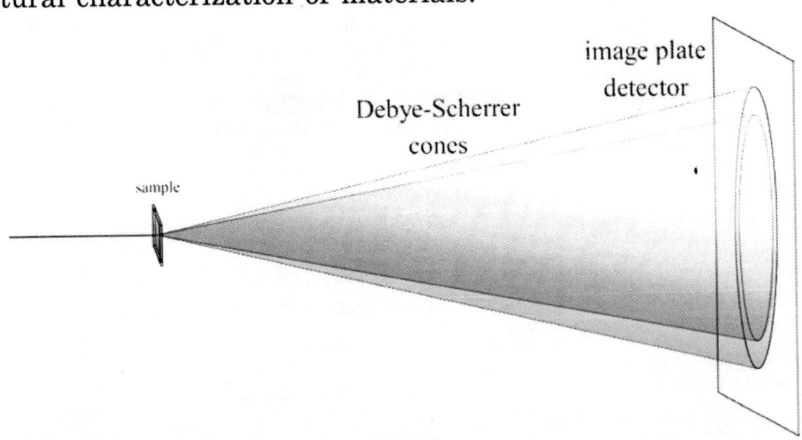

Figure: *Two-dimensional powder diffraction setup with flat plate detector.*

Ideally, every possible crystalline orientation is represented very equally in a powdered sample. The resulting orientational averaging causes the three-dimensional reciprocal space that is studied in single crystal diffraction to be projected onto a single dimension. The three-dimensional space can be described with (reciprocal) axes x^*, y^*, and z^* or alternatively in spherical coordinates q, φ^*, and χ^*. In powder diffraction, intensity is homogeneous over φ^* and χ^*, and only q remains as an important measurable quantity. In practice, it is sometimes necessary to rotate the sample orientation to eliminate the effects of texturing and achieve true randomness.

When the scattered radiation is collected on a flat plate detector, the rotational averaging leads to smooth diffraction rings around the beam axis, rather than the discrete Laue spots observed in single crystal diffraction. The angle between the beam axis and the ring is called the *scattering angle* and in X-ray crystallography always denoted as 2θ (in scattering of *visible* light the convention is usually to call it θ). In accordance with Bragg's law, each ring corresponds to a particular reciprocal lattice vector G in the sample crystal. This leads to the definition of the scattering vector as:

$$G = q = 2k\sin(\theta) = \frac{4\pi}{\lambda}\sin(\theta).$$

Powder diffraction data are usually presented as a diffractogram in which the diffracted intensity I is shown as function either of the scattering angle 2θ or as a function of the scattering vector q. The latter variable has the advantage that the diffractogram no longer depends on the value of the wavelength λ. The advent of synchrotron sources has widened the choice of wavelength considerably. To facilitate comparability of data obtained with different wavelengths the use of q is therefore recommended and gaining acceptability.

An instrument dedicated to performing powder measurements is called a powder diffractometre.

Uses

Relative to other methods of analysis, powder diffraction allows for rapid, non-destructive analysis of multi-component mixtures without the need for extensive sample preparation. This gives laboratories around the world the ability to quickly analyze unknown materials and perform materials characterization in such fields as metallurgy, mineralogy, forensic science, archeology, condensed matter physics, and the biological and pharmaceutical sciences. Identification is performed by comparison of the diffraction pattern to a known standard or to a database such as the International Centre for Diffraction Data's Powder Diffraction File (PDF) or the Cambridge Structural Database (CSD). Advances in hardware and software, particularly improved optics and fast detectors, have dramatically improved the analytical capability of the technique, especially relative to the speed of the analysis. The fundamental physics upon which the technique is based provides high precision and accuracy in the measurement of interplanar spacings, sometimes to fractions of an Ångström, resulting in authoritative identification frequently used in patents, criminal cases and other areas of law enforcement. The ability to analyze multiphase materials also allows analysis of how materials interact in a particular matrix such as a pharmaceutical tablet, a circuit board, a mechanical weld, a geologic core sampling, cement and concrete, or a pigment found in an historic painting. The method has been historically used for the identification and classification of minerals, but it can be used for any materials, even amorphous ones, so long as a suitable reference pattern is known or can be constructed.

Phase Identification

The most widespread use of powder diffraction is in the identification and characterization of crystalline solids, each of which produces a distinctive diffraction pattern. Both the positions (corresponding to lattice spacings) and the relative intensity of the lines are indicative of a particular phase and material, providing a "fingerprint" for comparison. A multi-phase mixture, e.g. a soil sample, will show more than one pattern superposed, allowing for determination of relative concentration.

J.D. Hanawalt, an analytical chemist who worked for Dow Chemical in the 1930s, was the first to realise the analytical potential of creating a database. Today it is represented by the Powder Diffraction File (PDF) of the International Centre for Diffraction Data (formerly Joint Committee for Powder Diffraction Studies). This has been made searchable by computer through the work of global software developers and equipment manufacturers. There are now over 550,000 reference materials in the 2006 Powder Diffraction File Databases, and these databases are interfaced to a wide variety of diffraction analysis software and distributed globally. The Powder Diffraction File contains many subfiles, such as minerals, metals and alloys, pharmaceuticals, forensics, excipients, superconductors, semiconductors, etc., with large collections of organic, organometallic and inorganic reference materials.

Crystallinity

In contrast to a crystalline pattern consisting of a series of sharp peaks, amorphous materials (liquids, glasses etc.) produce a broad background signal. Many polymers show semicrystalline behavior, *i.e.* part of the material forms an ordered crystallite by folding of the molecule. A single polymer molecule may well be folded into two different, adjacent crystallites and thus form a tie between the two. The tie part is prevented from crystallizing. The result is that the crystallinity will never reach 100%. Powder XRD can be used to determine the crystallinity by comparing the integrated intensity of the background pattern to that of the sharp peaks. Values obtained from powder XRD are typically comparable but not quite identical to those obtained from other methods such as DSC.

Lattice parameters

The position of a diffraction peak is 'independent' of the atomic positions within the cell and entirely determined by the size and shape of the unit cell of the crystalline phase. Each peak represents a certain lattice plane and can therefore be characterized by a Miller index. If the symmetry is high, e.g: cubic or hexagonal it is usually not too hard to identify the index of each peak, even for an unknown phase. This is particularly important in solid-state chemistry, where one is interested in finding and identifying new materials. Once a pattern has been indexed, this characterizes the reaction product and identifies it as a new solid phase. Indexing programs exist to deal with the harder cases, but if the unit cell is very large and the symmetry low (triclinic) success is not always guaranteed.

Expansion Tensors, Bulk Modulus

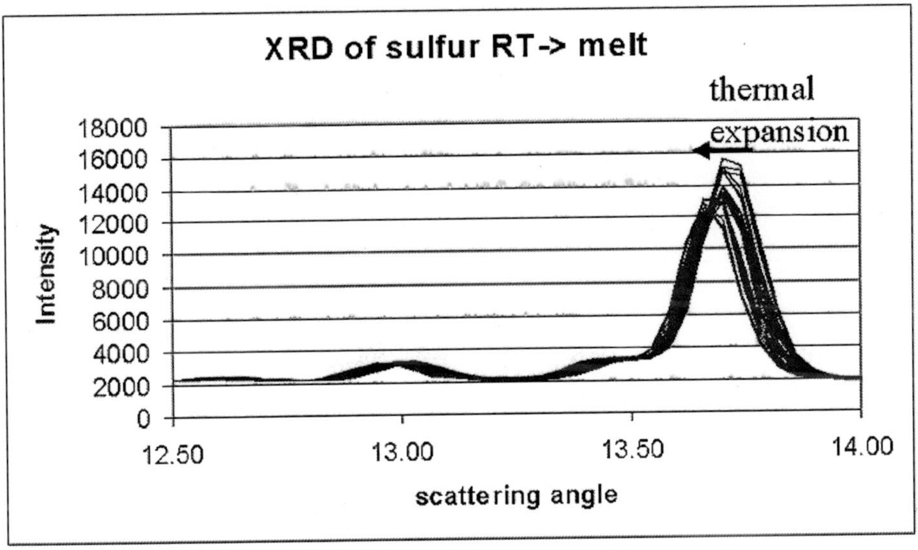

Figure: *Thermal expansion of a sulfur powder*

Cell parameters are somewhat temperature and pressure dependent. Powder diffraction can be combined with *in situ* temperature and pressure control. As these thermodynamic variables are changed, the observed diffraction peaks will migrate continuously to indicate higher or lower lattice spacings as the unit cell distorts. This allows for measurement of such quantities as the thermal expansion tensor and the isothermal bulk modulus, as well determination of the full equation of state of the material.

Phase Transitions

At some critical set of conditions, for example 0 °C for water at 1 atm, a new arrangement of atoms or molecules may become stable, leading to a phase transition. At this point new diffraction peaks will appear or old ones disappear according to the symmetry of the new phase. If the material melts to an isotropic liquid, all sharp lines will disappear and be replaced by a broad amorphous pattern. If the transition produces another crystalline phase, one set of lines will suddenly be replaced by another set. In some cases however lines will split or coalesce, e.g. if the material undergoes a continuous, second order phase transition. In such cases the symmetry may change because the existing structure is *distorted* rather than replaced by a completely different one. For example, the diffraction peaks for the lattice planes (100) and (001) can be found at two different values of q for a tetragonal phase, but if the symmetry becomes cubic the two peaks will come to coincide.

Crystal Structure Refinement and Determination

Crystal structure determination from powder diffraction data is extremely challenging due to the overlap of reflections in a powder experiment. A number of different methods exist for structural determination, such as simulated annealing and charge flipping. The crystal structures of known materials can be refined, i.e. as a function of temperature or pressure, using the Rietveld method. The Rietveld method is a so-called full pattern analysis technique. A crystal structure, together with instrumental and microstructural information, is used to generate a theoretical diffraction pattern that can be compared to the observed data. A least squares procedure is then used to minimize the difference between the calculated pattern and each point of the observed pattern by adjusting model parameters. Techniques to determine unknown structures from powder data do exist, but are somewhat specialised. A number of programs that can be used in structure determination are TOPAS, Fox, DASH, GSAS, EXPO2004, and a few others.

Size and Strain Broadening

There are many factors that determine the width B of a diffraction peak. These include:

1. instrumental factors
2. the presence of defects to the perfect lattice

3. differences in strain in different grains
4. the size of the crystallites

It is often possible to separate the effects of size and strain. Where size broadening is independent of q (K=1/d), strain broadening increases with increasing q-values. In most cases there will be both size and strain broadening. It is possible to separate these by combining the two equations in what is known as the Hall–Williamson method:

$$B\cdot\cos(\theta) = \frac{k\lambda}{D} + \eta\cdot\sin(\theta),$$

Thus, when we plot $B\cdot\cos(\theta)$ vs. $\sin(\theta)$ we get a straight line with slope η and intercept $\frac{k\lambda}{D}$.

The expression is a combination of the Scherrer equation for size broadening and the Stokes and Wilson expression for strain broadening. The value of η is the strain in the crystallites, the value of D represents the size of the crystallites. The constant k is typically close to unity and ranges from 0.8 to 1.39.

Comparison of X-Ray and Neutron Scattering

X-ray photons scatter by interaction with the electron cloud of the material, neutrons are scattered by the nuclei. This means that, in the presence of heavy atoms with many electrons, it may be difficult to detect light atoms by X-ray diffraction.

In contrast, the neutron scattering lengths of most atoms are approximately equal in magnitude. Neutron diffraction techniques may therefore be used to detect light elements such as oxygen or hydrogen in combination with heavy atoms. The neutron diffraction technique therefore has obvious applications to problems such as determining oxygen displacements in materials like high temperature superconductors and ferroelectrics, or to hydrogen bonding in biological systems.

A further complication in the case of neutron scattering from hydrogenous materials is the strong incoherent scattering of hydrogen (80.27(6) barn). This leads to a very high background in neutron diffraction experiments, and may make structural investigations impossible. A common solution is deuteration, i.e., replacing the 1-H atoms in the sample with deuterium (2-H). The incoherent scattering length of deuterium is much smaller (2.05(3) barn) making structural investigations significantly easier. However, in some systems, replacing hydrogen with deuterium may alter the structural and dynamic properties of interest.

As neutrons also have a magnetic moment, they are additionally scattered by any magnetic moments in a sample. In the case of long range magnetic order, this leads to the appearance of new Bragg reflections. In most simple cases, powder diffraction may be used to determine the size of the moments and their spatial orientation.

Aperiodically-Arranged Clusters

Predicting the scattered intensity in powder diffraction patterns from gases, liquids, and randomly distributed nano-clusters in the solid state is (to first order) done rather elegantly with the Debye scattering equation:

$$I(q) = \sum_{i=1}^{N} \sum_{j=1}^{N} f_i(q) f_j(q) \frac{\sin(qr_{ij})}{qr_{ij}},$$

where the magnitude of the scattering vector q is in reciprocal lattice distance units, N is the number of atoms, $f_i(q)$ is the atomic scattering factor for atom i and scattering vector q, while r_{ij} is the distance between atom i and atom j. One can also use this to predict the effect of nano-crystallite shape on detected diffraction peaks, even if in some directions the cluster is only one atom thick.

Devices

Cameras: The simplest cameras for X-ray powder diffraction consist of a small capillary and either a flat plate detector (originally a piece of X-ray film, now more and more a flat-plate detector or a CCD-camera) or a cylindrical one (originally a piece of film in a cookie-jar, but increasingly bent position sensitive detectors are used). The two types of cameras are known as the Laue and the Debye–Scherrer camera.

In order to ensure complete powder averaging, the capillary is usually spun around its axis. For neutron diffraction vanadium cylinders are used as sample holders. Vanadium has a negligible absorption and coherent scattering cross section for neutrons and is hence nearly invisible in a powder diffraction experiment. Vanadium does however have a considerable incoherent scattering cross section which may cause problems for more sensitive techniques such as neutron inelastic scattering.

A later development in X-ray cameras is the Guinier camera. It is built around a *focusing* bent crystal monochromator. The sample is usually placed in the focusing beam, e.g. as a dusting on a piece of sticky tape. A cylindrical piece of film (or electronic multichannel detector) is put on the focusing circle, but the incident beam prevented from reaching the detector to prevent damage from its high intensity.

Diffractometres: Diffractometres can be operated both in transmission and in reflection configurations. The reflection one is more common. The powder sample is filled in a small disc like container and its surface carefully flattened. The disc is put on one axis of the diffractometre and tilted by an angle θ while a detector (scintillation counter) rotates around it on an arm at twice this angle. This configuration is known under the name Bragg–Brentano theta-2 theta.

Another configuration is the Bragg–Brentano theta-theta configuration in which the sample is stationary while the X-ray tube and the detector are rotated around it. The angle formed between the tube and the detector is 2theta. This configuration is most convenient for loose powders.

Position sensitive and area detectors, which allow collection of multiple angles at once, are becoming more popular on currently supplied instrumentation.

Neutron Diffraction

Sources that produce a neutron beam of suitable intensity and speed for diffraction are only available at a small number of research reactors and spallation sources in the world. Angle dispersive (fixed wavelength) instruments typically have a battery of individual detectors arranged in a cylindrical fashion around the sample holder, and can therefore collect scattered intensity simultaneously on a large 2θ range. Time of flight instruments normally have a small range of banks at different scattering angles which collect data at varying resolutions.

X-Ray Tubes

Laboratory X-ray diffraction equipment relies on the use of an X-ray tube, which is used to produce the X-rays.

The most commonly used laboratory X-ray tube uses a copper anode, but cobalt and molybdenum are also popular. The wavelength in nm varies for each source. The table below shows these wavelengths, determined by Bearden and quoted in the International Tables for X-ray Crystallography (all values in nm):

Element	***Ka (weight average)***	***Ka2 (strong)***	***Ka1 (very strong)***	***Kβ (weak)***
Cr	0.229100	0.229361	0.228970	0.208487
Fe	0.193736	0.193998	0.193604	0.175661
Co	0.179026	0.179285	0.178897	0.162079
Cu	0.154184	0.154439	0.154056	0.139222
Mo	0.071073	0.071359	0.070930	0.063229

According to the last re-examination of Holzer et al. (1997), these values are respectively:

Element	*Ka2*	*Ka1*	*Kβ*
Cr	0.2293663	0.2289760	0.2084920
Co	0.1792900	0.1789010	0.1620830
Cu	0.1544426	0.1540598	0.1392250
Mo	0.0713609	0.0709319	0.0632305

Other Sources

In-house applications of X-ray diffraction has always been limited to the relatively few wavelengths shown in the table above. The available choice was much needed because the combination of certain wavelengths and certain elements present in a sample can lead to strong fluorescence which increases the background in the diffraction pattern. A notorious example is the presence of iron in a sample when using copper radiation. In general elements just below the anode element in the period system need to be avoided.

Another limitation is that the intensity of traditional generators is relatively low, requiring lengthy exposure times and precluding any time dependent measurement. The advent of synchrotron sources has drastically changed this picture and caused powder diffraction methods to enter a whole new phase of development. Not only is there a much wider choice of wavelengths available, the high brilliance of the synchrotron radiation makes it possible to observe changes in the pattern during chemical reactions, temperature ramps, changes in pressure and the like.

The tunability of the wavelength also makes it possible to observe anomalous scattering effects when the wavelength is chosen close to the absorption edge of one of the elements of the sample.

Neutron diffraction has never been an in house technique because it requires the availability of an intense neutron beam only available at a nuclear reactor or spallation source. Typically the available neutron flux, and the weak interaction between neutrons and matter, require relative large samples.

Advantages and Disadvantages

Although it possible to solve crystal structures from powder X-ray data alone, its single crystal analogue is a far more powerful technique for structure determination. This is directly related to the fact that much information is lost by the collapse of the 3D space onto a 1D axis. Nevertheless powder X-ray diffraction is a powerful and useful technique in its own right. It is mostly used to characterize and identify *phases* rather than solving structures.

The great advantages of the technique are:

- simplicity of sample preparation
- rapidity of measurement
- the ability to analyze mixed phases, e.g. soil samples
- "in situ" structure determination

By contrast growth and mounting of large single crystals is notoriously difficult. In fact there are many materials for which despite many attempts it has not proven possible to obtain single crystals. Many materials are readily available with sufficient microcrystallinity for powder diffraction, or samples may be easily ground from larger crystals. In the field of solid-state chemistry that often aims at synthesizing *new* materials, single crystals thereof are typically not immediately available. Powder diffraction is therefore one of the most powerful methods to identify and characterize new materials in this field.

Particularly for neutron diffraction, which requires larger samples than X-ray diffraction due to a relatively weak scattering cross section, the ability to use large samples can be critical, although new more brilliant neutron sources are being built that may change this picture.

Since all possible crystal orientations are measured simultaneously, collection times can be quite short even for small and weakly scattering samples. This is not merely convenient, but can be essential for samples which are unstable either inherently or under X-ray or neutron bombardment, or for time-resolved studies. For the latter it is desirable to have a strong radiation source. The advent of synchrotron radiation and modern neutron sources has therefore done much to revitalize the powder diffraction field because it is now possible to study temperature dependent changes, reaction kinetics and so forth by means of time dependent powder diffraction.

Bragg Diffraction

In physics, Bragg's law gives the angles for coherent and incoherent scattering from a crystal lattice. When X-rays are incident on an atom, they make the electronic cloud move as does any electromagnetic wave. The movement of these charges re-radiates waves with the same frequency (blurred slightly due to a variety of effects); this phenomenon is known as Rayleigh scattering (or elastic scattering). The scattered waves can themselves be scattered but this secondary scattering is assumed to be negligible.

A similar process occurs upon scattering neutron waves from the nuclei or by a coherent spin interaction with an unpaired electron. These re-emitted wave fields interfere with each other either constructively or destructively (overlapping waves either add together to produce stronger peaks or subtract from each other to some degree), producing a diffraction pattern on a detector or film. The resulting wave interference pattern is the basis of diffraction analysis. This analysis is called *Bragg diffraction*.

Bragg diffraction (also referred to as the Bragg formulation of X-ray diffraction) was first proposed by William Lawrence Bragg and William Henry Bragg in 1913 in response to their discovery that crystalline solids produced surprising patterns of reflected X-rays (in contrast to that of, say, a liquid). They found that these crystals, at certain specific wavelengths and incident angles, produced intense peaks of reflected radiation (known as *Bragg peaks*). The concept of Bragg diffraction applies equally to neutron diffraction and electron diffraction processes. Both neutron and X-ray wavelengths are comparable with inter-atomic distances (~150 pm) and thus are an excellent probe for this length scale.

Figure: *X-rays interact with the atoms in a crystal.*

W. L. Bragg explained this result by modelling the crystal as a set of discrete parallel planes separated by a constant parameter d. It was proposed that the incident X-ray radiation would produce a Bragg peak if their reflections off the various planes interfered constructively. The interference is constructive when the phase shift is a multiple of 2π; this condition can be expressed by Bragg's law:

$$n\lambda = 2d \sin\theta,$$

where n is an integer, λ is the wavelength of incident wave, d is the spacing between the planes in the atomic lattice, and θ is the angle between the incident ray and the scattering planes. Note that moving particles, including electrons, protons and neutrons, have an associated De Broglie wavelength.

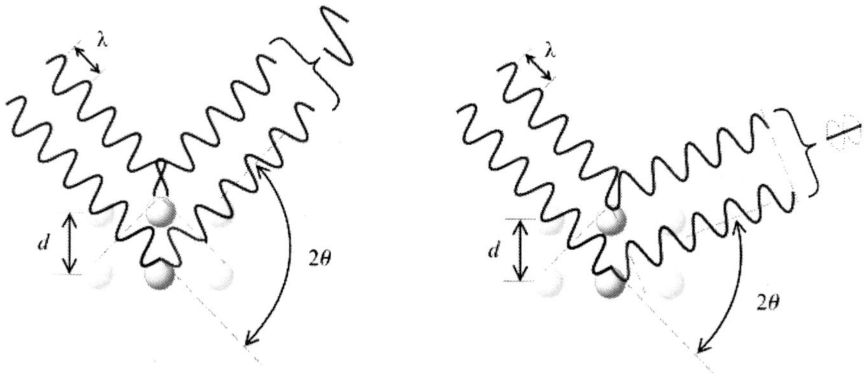

***Figure:** According to the 2è deviation, the phase shift causes constructive (left figure) or destructive (right figure) interferences.*

Bragg's Law was derived by physicist Sir William Lawrence Bragg in 1912 and first presented on 11 November 1912 to the Cambridge Philosophical Society. Although simple, Bragg's law confirmed the existence of real particles at the atomic scale, as well as providing a powerful new tool for studying crystals in the form of X-ray and neutron diffraction. William Lawrence Bragg and his father, Sir William Henry Bragg, were awarded the Nobel Prize in physics in 1915 for their work in determining crystal structures beginning with NaCl, ZnS, and diamond. They are the only father-son team to jointly win. W. L. Bragg was 25 years old, making him the youngest Nobel laureate.

Bragg Condition

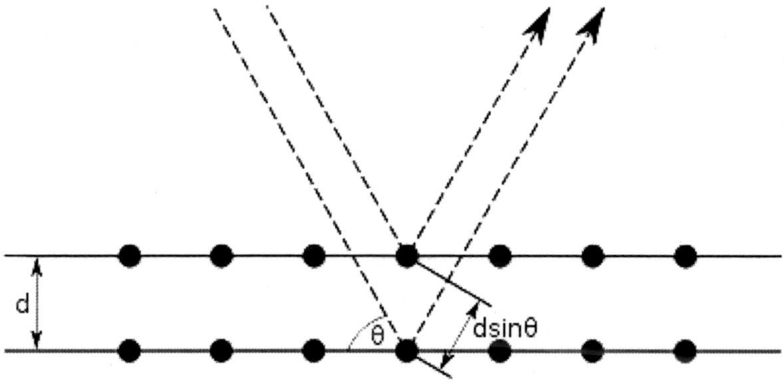

***Figure:** Bragg diffraction. Two beams with identical wavelength and phase approach a crystalline solid and are scattered off two different atoms within it. The lower beam traverses an extra length of 2dsinθ. Constructive interference occurs when this length is equal to an integer multiple of the wavelength of the radiation.*

Bragg diffraction occurs when electromagnetic radiation or subatomic particle waves with wavelength comparable to atomic spacings are incident upon a crystalline sample, are scattered in a specular fashion by the atoms in the system, and undergo constructive interference in accordance to Bragg's law. For a crystalline solid, the waves are scattered from lattice planes separated by the interplanar distance *d*. Where the scattered waves interfere constructively, they remain in phase since the path length of each wave is equal to an integer multiple of the wavelength. The path difference between two waves undergoing constructive interference is given by $2d\sin\theta$, where θ is the scattering angle. This leads to Bragg's law, which describes the condition for constructive interference from successive crystallographic planes (*h*, *k*, and *l*, as given in Miller Notation) of the crystalline lattice:

$$2d\sin\theta = n\lambda,$$

where *n* is an integer determined by the order given, and λ is the wavelength. A diffraction pattern is obtained by measuring the intensity of scattered waves as a function of scattering angle. Very strong intensities known as Bragg peaks are obtained in the diffraction pattern when scattered waves satisfy the Bragg condition.

It should be taken into account that if only two planes of atoms were diffracting, as shown in the pictures, then the transition from constructive to destructive interference would be gradual as the angle is varied. However, since many atomic planes are interfering in real materials, very sharp peaks surrounded by mostly destructive interference result.

Reciprocal Space

Although the misleading common opinion reigns that Bragg's law measures atomic distances in real space, it does not. This first statement only seems to be true if it's further elaborated that distances measured during a Bragg experiment are inversely proportional to the distance d in the lattice diagram. Furthermore, the $n\lambda$ term demonstrates that it measures the number of wavelengths fitting between two rows of atoms, thus measuring reciprocal distances. Reciprocal lattice vectors describe the set of lattice planes as a normal vector to this set with length $G = 2\pi / d$. Max von Laue had interpreted this correctly in a vector form, the Laue equation

$$\vec{G} = \vec{k}_f - \vec{k}_i,$$

where $\vec{G}$ is a reciprocal lattice vector and $\vec{k}_f$ and $\vec{k}_i$ are the wave vectors of the diffracted and the incident beams respectively.

Together with the condition for elastic scattering $|k_f| = |k_i|$ and the introduction of the scattering angle 2θ this leads equivalently to Bragg's equation. This is simply explained by the conservation of momentum. In this system the scanning variable can be the length or the direction of the incident or exit wave vectors relating to energy- and angle-dispersive setups. The simple relationship between diffraction angle and Q-space is then

$$Q = \frac{4\pi \sin(\theta)}{\lambda}.$$

The concept of reciprocal lattice is the Fourier space of a crystal lattice and necessary for a full mathematical description of wave mechanics.

Alternate Derivation

Suppose that a single monochromatic wave (of any type) is incident on aligned planes of lattice points, with separation *d*, at angle θ. Points A and C are on one plane, and B is on the plane below. Points ABCC' form a quadrilateral.

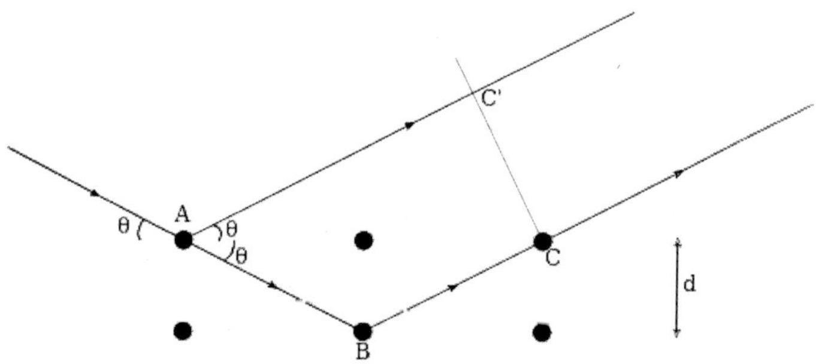

There will be a path difference between the ray that gets reflected along AC' and the ray that gets transmitted, then reflected, along AB and BC respectively. This path difference is

$$(AB + BC) - (AC').$$

The two separate waves will arrive at a point with the same phase, and hence undergo constructive interference, if and only if this path difference is equal to any integer value of the wavelength, i.e.

$$(AB + BC) - (AC') = n\lambda,$$

where the same definition of n and λ apply as above.

Therefore,

$$AB = BC = \frac{d}{\sin\theta} \text{ and } AC = \frac{2d}{\tan\theta},$$

from which it follows that

$$AC' = AC \cdot \cos\theta = \frac{2d}{\tan\theta}\cos\theta = \left(\frac{2d}{\sin\theta}\cos\theta\right)\cos\theta = \frac{2d}{\sin\theta}\cos^2\theta.$$

Putting everything together,

$$n\lambda = \frac{2d}{\sin\theta}(1-\cos^2\theta) = \frac{2d}{\sin\theta}\sin^2\theta,$$

which simplifies to

$$n\lambda = 2d\sin\theta,$$

which is Bragg's law.

Bragg Scattering of Visible Light by Colloids

A colloidal crystal is a highly ordered array of particles which can be formed over a very long range (from a few millimetres to one centimetre) in length, and which appear analogous to their atomic or molecular counterparts. The periodic arrays of spherical particles make similar arrays of interstitial voids (the spaces between the particles), which act as a natural diffraction grating for visible light waves, especially when the interstitial spacing is of the same order of magnitude as the incident lightwave.

Thus, it has been known for many years that, due to repulsive Coulombic interactions, electrically charged macromolecules in an aqueous environment can exhibit long-range crystal-like correlations with interparticle separation distances often being considerably greater than the individual particle diameter.

In all of these cases in nature, the same brilliant iridescence (or play of colours) can be attributed to the diffraction and constructive interference of visible lightwaves which satisfy Bragg's law, in a matter analogous to the scattering of X-rays in crystalline solid.

Selection Rules and Practical Crystallography

Bragg's law, as stated above, can be used to obtain the lattice spacing of a particular cubic system through the following relation:

$$d = \frac{a}{\sqrt{h^2+k^2+l^2}},$$

where a is the lattice spacing of the cubic crystal, and h, k, and l are the Miller indices of the Bragg plane. Combining this relation with Bragg's law:

$$\left(\frac{\lambda}{2a}\right)^2 = \frac{\sin^2\theta}{h^2+k^2+l^2}.$$

One can derive selection rules for the Miller indices for different cubic Bravais lattices; here, selection rules for several will be given as is.

Bravais lattice	***Example compounds***	***Allowed reflections***	***Forbidden reflections***
Simple cubic	Po	Any h, k, l	None
Body-centered cubic	Fe, W, Ta, Cr	$h + k + l$ = even	$h + k + l$ = odd
Face-centered cubic	Cu, Al, Ni, NaCl, LiH, PbS	h, k, l all odd or all even	h, k, l mixed odd and even
Diamond F.C.C.	Si, Ge	all odd, or all even with $h+k+l = 4n$	h, k, l mixed odd and even, or all even with $h+k+l \neq 4n$
Triangular lattice	Ti, Zr, Cd, Be	l even, $h + 2k \neq 3n$	$h + 2k = 3n$ for odd l

These selection rules can be used for any crystal with the given crystal structure. KCl exhibits a fcc cubic structure.

However, the K^+ and the Cl^- ion have the same number of electrons and are quite close in size, so that the diffraction pattern becomes essentially the same as for a simple cubic structure with half the lattice parameter. Selection rules for other structures can be referenced elsewhere, or derived.

3

Condensed Matter Physics

Condensed matter physics is a branch of physics that deals with the physical properties of condensed phases of matter. Condensed matter physicists seek to understand the behavior of these phases by using physical laws. In particular, these include the laws of quantum mechanics, electromagnetism and statistical mechanics.

The most familiar condensed phases are solids and liquids, while more exotic condensed phases include the superconducting phase exhibited by certain materials at low temperature, the ferromagnetic and antiferromagnetic phases of spins on atomic lattices, and the Bose–Einstein condensate found in cold atomic systems. The study of condensed matter physics involves measuring various material properties via experimental probes along with using techniques of theoretical physics to develop mathematical models that help in understanding physical behavior.

The diversity of systems and phenomena available for study makes condensed matter physics the most active field of contemporary physics: one third of all American physicists identify themselves as condensed matter physicists, and The Division of Condensed Matter Physics (DCMP) is the largest division of the American Physical Society. The field overlaps with chemistry, materials science, and nanotechnology, and relates closely to atomic physics and biophysics. Theoretical condensed matter physics shares important concepts and techniques with theoretical particle and nuclear physics.

A variety of topics in physics such as crystallography, metallurgy, elasticity, magnetism, etc., were treated as distinct areas, until the 1940s when they were grouped together as *Solid state physics*. Around the 1960s, the study of physical properties of liquids was added to this list, and it came to be known as condensed matter physics. According to physicist Phil Anderson, the term was coined by him and Volker Heine when they changed the name of their group at the Cavendish Laboratories, Cambridge from "Solid state theory" to "Theory of Condensed Matter", as they felt it did not exclude their interests in the study of liquids, nuclear matter and so on. The Bell Labs (then known as the *Bell Telephone Laboratories*) was one of the first institutes to conduct a research program in condensed matter physics.

References to "condensed" state can be traced to earlier sources. For example, in the introduction to his 1947 "Kinetic theory of liquids" book, Yakov Frenkel proposed that "The kinetic theory of liquids must accordingly be developed as a generalization and extension of the kinetic theory of solid bodies. As a matter of fact, it would be more correct to unify them under the title of "condensed bodies".

One of the first studies of condensed states of matter was by English chemist Humphry Davy, in the first decades of the 19th century. Davy observed that of the 40 chemical elements known at the time, 26 had metallic properties such as lustre, ductility and high electrical and thermal conductivity. This indicated that the atoms in Dalton's atomic theory were not indivisible as Dalton claimed, but had inner structure. Davy further claimed that elements that were then believed to be gases, such as nitrogen and hydrogen could be liquified under the right conditions and would then behave as metals.

In 1823, Michael Faraday, then an assistant in Davy's lab, successfully liquified chlorine and went on to liquify all known gaseous elements, with the exception of nitrogen, hydrogen and oxygen. Shortly after, in 1869, Irish chemist Thomas Andrews studied the phase transition from a liquid to a gas and coined the term critical point to describe the condition where a gas and a liquid were indistinguishable as phases, and Dutch physicist Johannes van der Waals supplied the theoretical framework which allowed the prediction of critical behavior based on measurements at much higher temperatures. By 1908, James Dewar and H. Kamerlingh Onnes were successfully able to liquify hydrogen and then newly discovered helium, respectively.

Paul Drude proposed the first theoretical model for a classical electron moving through a metallic solid. Drude's model described properties of metals in terms of a gas of free electrons, and was the first microscopic model to explain empirical observations such as the Wiedemann–Franz law. However, despite the success of Drude's free electron model, it had one notable problem, in that it was unable to correctly explain the electronic contribution to the specific heat of metals, as well as the temperature dependence of resistivity at low temperatures.

In 1911, just three years after helium was first liquified, Onnes working at University of Leiden discovered superconductivity in mercury, when he observed the electrical resistivity in mercury to vanish when the temperature was lowered below a certain value. The phenomenon completely surprised the best theoretical physicists of the time, and it remained unexplained for several decades. Albert Einstein, in 1922, said regarding contemporary theories of superconductivity that "with our far-reaching ignorance of the quantum mechanics of composite systems we are very far from being able to compose a theory out of these vague ideas".

Advent of Quantum Mechanics

Figure: *A replica of the first point-contact transistor in Bell labs*

Drude's classical model was augmented by Felix Bloch, Arnold Sommerfeld, and independently by Wolfgang Pauli, who used quantum mechanics to describe the motion of a quantum electron in a periodic lattice. In particular, Sommerfeld's theory accounted for the Fermi–

Dirac statistics satisfied by electrons and was better able to explain the heat capacity and resistivity. The structure of crystalline solids was studied by Max von Laue and Paul Knipping, when they observed the X-ray diffraction pattern of crystals, and concluded that crystals get their structure from periodic lattices of atoms.

The mathematics of crystal structures developed by Auguste Bravais, Yevgraf Fyodorov and others was used to classify crystals by their symmetry group, and tables of crystal structures were the basis for the series *International Tables of Crystallography*, first published in 1935. Band structure calculations was first used in 1930 to predict the properties of new materials, and in 1947 John Bardeen, Walter Brattain and William Shockley developed the first semiconductor-based transistor, heralding a revolution in electronics.

In 1879, Edwin Herbert Hall working at the Johns Hopkins University discovered the development of a voltage across conductors transverse to an electric current in the conductor and magnetic field perpendicular to the current. This phenomenon arising due to the nature of charge carriers in the conductor came to be known as the Hall effect, but it was not properly explained at the time, since the electron was experimentally discovered 18 years later. After the advent of quantum mechanics, Lev Landau in 1930 predicted the quantization of the Hall conductance for electrons confined to two dimensions.

Magnetism as a property of matter has been known since pre-historic times. However, the first modern studies of magnetism only started with the development of electrodynamics by Faraday, Maxwell and others in the nineteenth century, which included the classification of materials as ferromagnetic, paramagnetic and diamagnetic based on their response to magnetization. Pierre Curie studied the dependence of magnetization on temperature and discovered the Curie point phase transition in ferromagnetic materials. In 1906, Pierre Weiss introduced the concept of magnetic domains to explain the main properties of ferromagnets.

The first attempt at a microscopic description of magnetism was by Wilhelm Lenz and Ernst Ising through the Ising model that described magnetic materials as consisting of a periodic lattice of spins that collectively acquired magnetization. The Ising model was solved exactly to show that spontaneous magnetization cannot occur in one dimension but is possible in higher-dimensional lattices. Further research such as by Bloch on spin waves and Néel on antiferromagnetism led to the development of new magnetic materials with applications to magnetic storage devices.

Modern Many Body Physics

The Sommerfeld model and spin models for ferromagnetism illustrated the successful application of quantum mechanics to condensed matter problems in the 1930s. However, there still were several unsolved problems, most notably the description of superconductivity and the Kondo effect. After World War II, several ideas from quantum field theory were applied to condensed matter problems. These included recognition of collective modes of excitation of solids and the important notion of a quasiparticle. Russian physicist Lev Landau used the idea for the Fermi liquid theory wherein low energy properties of interacting fermion systems were given in terms of what are now known as Landau-quasiparticles. Landau also developed a mean field theory for continuous phase transitions, which described ordered phases as spontaneous breakdown of symmetry. The theory also introduced the notion of an order parameter to distinguish between ordered phases. Eventually in 1965, John Bardeen, Leon Cooper and John Schrieffer developed the so-called BCS theory of superconductivity, based on the discovery that arbitrarily small attraction between two electrons can give rise to a bound state called a Cooper pair.

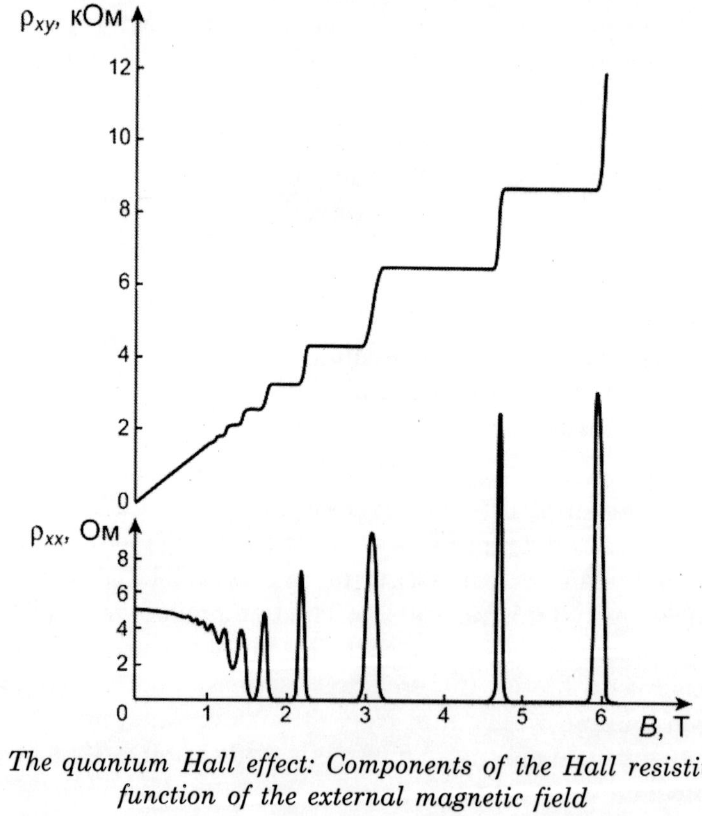

Figure: *The quantum Hall effect: Components of the Hall resistivity as a function of the external magnetic field*

The study of phase transition and the critical behavior of observables, known as critical phenomena, was a major field of interest in the 1960s. Leo Kadanoff, Benjamin Widom and Michael Fisher developed the ideas of critical exponents and scaling. These ideas were unified by Kenneth Wilson in 1972, under the formalism of the renormalization group in the context of quantum field theory.

The quantum Hall effect was discovered by Klaus von Klitzing in 1980 when he observed the Hall conductivity to be integer multiples of a fundamental constant. The effect was observed to be independent of parameters such as the system size and impurities, and in 1981, theorist Robert Laughlin proposed a theory describing the integer states in terms of a topological invariant called the Chern number. Shortly after, in 1982, Horst Störmer and Daniel Tsui observed the fractional quantum Hall effect where the conductivity was now a rational multiple of a constant. Laughlin, in 1983, realised that this was a consequence of quasiparticle interaction in the Hall states and formulated a variational solution, known as the Laughlin wavefunction. The study of topological properties of the fractional Hall effect remains an active field of research.

In 1987, Karl Müller and Johannes Bednorz discovered the first high temperature superconductor, a material which was superconducting at temperatures as high as 50 Kelvin. It was realised that the high temperature superconductors are examples of strongly correlated materials where the electron–electron interactions play an important role. A satisfactory theoretical description of high-temperature superconductors is still not known and the field of strongly correlated materials continues to be an active research topic.

In 2009, David Field and researchers at Aarhus University discovered spontaneous electric fields when creating prosaic films of various gases. This has more recently expanded to form the research area of spontelectrics.

Theoretical

Theoretical condensed matter physics involves the use of theoretical models to understand properties of states of matter. These include models to study the electronic properties of solids, such as the Drude model, the Band structure and the density functional theory. Theoretical models have also been developed to study the physics of phase transitions, such as the Ginzburg–Landau theory, critical exponents and the use of mathematical techniques of quantum field theory and the renormalization group. Modern theoretical studies involve the use of numerical computation of electronic structure and mathematical tools to understand phenomena such as high-temperature superconductivity, topological phases and gauge symmetries.

Emergence

Theoretical understanding of condensed matter physics is closely related to the notion of emergence, wherein complex assemblies of particles behave in ways dramatically different from their individual constituents. For example, a range of phenomena related to high temperature superconductivity are not well understood, although the microscopic physics of individual electrons and lattices is well known. Similarly, models of condensed matter systems have been studied where collective excitations behave like photons and electrons, thereby describing electromagnetism as an emergent phenomenon. Emergent properties can also occur at the interface between materials: one example is the lanthanum-aluminate-strontium-titanate interface, where two non-magnetic insulators are joined to create conductivity, superconductivity, and ferromagnetism.

Electronic Theory of Solids

The metallic state has historically been an important building block for studying properties of solids. The first theoretical description of metals was given by Paul Drude in 1900 with the Drude model, which explained electrical and thermal properties by describing a metal as an ideal gas of then-newly discovered electrons. This classical model was then improved by Arnold Sommerfeld who incorporated the Fermi–Dirac statistics of electrons and was able to explain the anomalous behavior of the specific heat of metals in the Wiedemann–Franz law. In 1913, X-ray diffraction experiments revealed that metals possess periodic lattice structure. Swiss physicist Felix Bloch provided a wave function solution to the Schrödinger equation with a periodic potential, called the Bloch wave.

Calculating electronic properties of metals by solving the many-body wavefunction is often computationally hard, and hence, approximation techniques are necessary to obtain meaningful predictions. The Thomas–Fermi theory, developed in the 1920s, was used to estimate electronic energy levels by treating the local electron density as a variational parameter. Later in the 1930s, Douglas Hartree, Vladimir Fock and John Slater developed the so-called Hartree–Fock wavefunction as an improvement over the Thomas–Fermi model. The Hartree–Fock method accounted for exchange statistics of single particle electron wavefunctions, but not for their Coulomb interaction. Finally in 1964–65, Walter Kohn, Pierre Hohenberg and Lu Jeu Sham proposed the density functional theory which gave realistic descriptions for bulk and surface properties of metals. The density functional theory (DFT) has been widely used since the 1940s for band structure calculations of variety of solids.

Symmetry Breaking

Certain states of matter exhibit symmetry breaking, where the relevant laws of physics possess some symmetry that is broken. A common example is crystalline solids, which break continuous translational symmetry. Other examples include magnetized ferromagnets, which break rotational symmetry, and more exotic states such as the ground state of a BCS superconductor, that breaks U(1) rotational symmetry.

Goldstone's theorem in quantum field theory states that in a system with broken continuous symmetry, there may exist excitations with arbitrarily low energy, called the Goldstone bosons. For example, in crystalline solids, these correspond to phonons, which are quantized versions of lattice vibrations.

Phase Transition

The study of critical phenomena and phase transitions is an important part of modern condensed matter physics. Phase transition refers to the change of phase of a system, which is brought about by change in an external parameter such as temperature. In particular, quantum phase transitions refer to transitions where the temperature is set to zero, and the phases of the system refer to distinct ground states of the Hamiltonian. Systems undergoing phase transition display critical behavior, wherein several of their properties such as correlation length, specific heat and susceptibility diverge. Continuous phase transitions are described by the Ginzburg–Landau theory, which works in the so-called mean field approximation. However, several important phase transitions, such as the Mott insulator–superfluid transition, are known that do not follow the Ginzburg–Landau paradigm. The study of phase transitions in strongly correlated systems is an active area of research.

Experimental

Experimental condensed matter physics involves the use of experimental probes to try to discover new properties of materials. Experimental probes include effects of electric and magnetic fields, measurement of response functions, transport properties and thermometry. Commonly used experimental techniques include spectroscopy, with probes such as X-rays, infrared light and inelastic neutron scattering; study of thermal response, such as specific heat and measurement of transport via thermal and heat conduction.

Figure: *Image of X-ray diffraction pattern from a protein crystal.*

Scattering

Several condensed matter experiments involve scattering of an experimental probe, such as X-ray, optical photons, neutrons, etc., on constituents of a material. The choice of scattering probe depends on the observation energy scale of interest. Visible light has energy on the scale of 1 eV and is used as a scattering probe to measure variations in material properties such as dielectric constant and refractive index. X-rays have energies of the order of 10 keV and hence are able to probe atomic length scales, and are used to measure variations in electron charge density. Neutrons can also probe atomic length scales and are used to study scattering off nuclei and electron spins and magnetization (as neutrons themselves have spin but no charge). Coulomb and Mott scattering measurements can be made by using electron beams as scattering probes, and similarly, positron annihilation can be used as an indirect measurement of local electron density. Laser spectroscopy is used as a tool for studying phenomena with energy in the range of visible light, for example, to study non-linear optics and forbidden transitions in media.

External Magnetic Fields

In experimental condensed matter physics, external magnetic fields act as thermodynamic variables that control the state, phase transitions and properties of material systems. Nuclear magnetic resonance (NMR) is a technique by which external magnetic fields can be used to find resonance modes of individual electrons, thus giving information about the atomic, molecular and bond structure of their neighbourhood. NMR experiments can be made in magnetic fields with strengths up to 65 Tesla. Quantum oscillations is another experimental technique where high magnetic fields are used to study material properties such as the geometry of the Fermi surface. The quantum hall effect is another example of measurements with high magnetic fields where topological properties such as Chern–Simons angle can be measured experimentally.

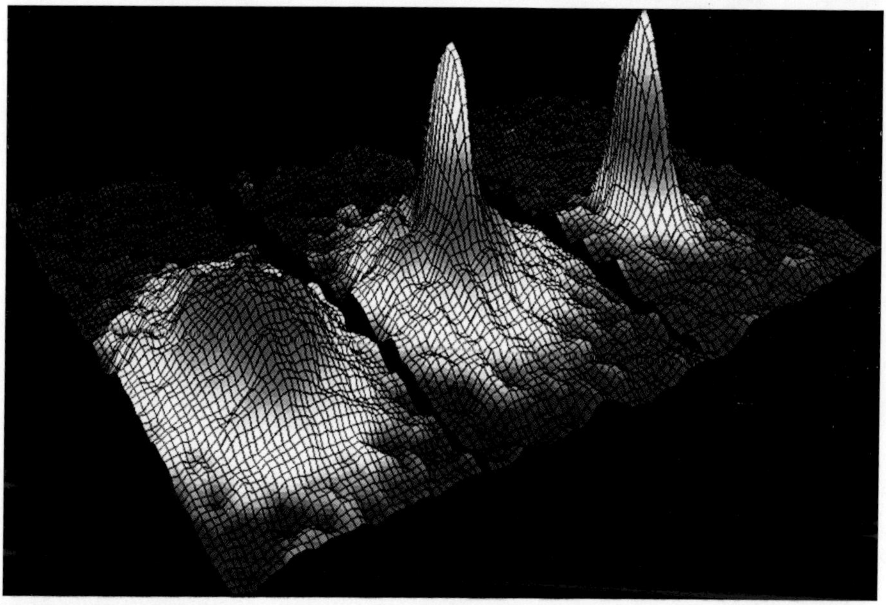

Figure: *The first Bose–Einstein condensate observed in a gas of ultracold rubidium atoms. The blue and white areas represent higher density.*

Cold Atomic Gases

Cold ion trapping in optical lattices is an experimental tool commonly used in condensed matter as well as atomic, molecular, and optical physics. The technique involves using optical lasers to create an interference pattern, which acts as a "lattice", in which ions or atoms can be placed at very low temperatures. Cold atoms in optical lattices are used as "quantum simulators", that is, they act as controllable systems that can model behavior of more complicated systems, such as frustrated magnets. In particular, they are used to engineer one-, two- and three-dimensional lattices for a Hubbard model with pre-specified parameters. and to study phase transitions for Néel and spin liquid ordering.

In 1995, a gas of rubidium atoms cooled down to a temperature of 170 nK was used to experimentally realise the Bose–Einstein condensate, a novel state of matter originally predicted by S. N. Bose and Albert Einstein, wherein a large number of atoms occupy a single quantum state.

Applications

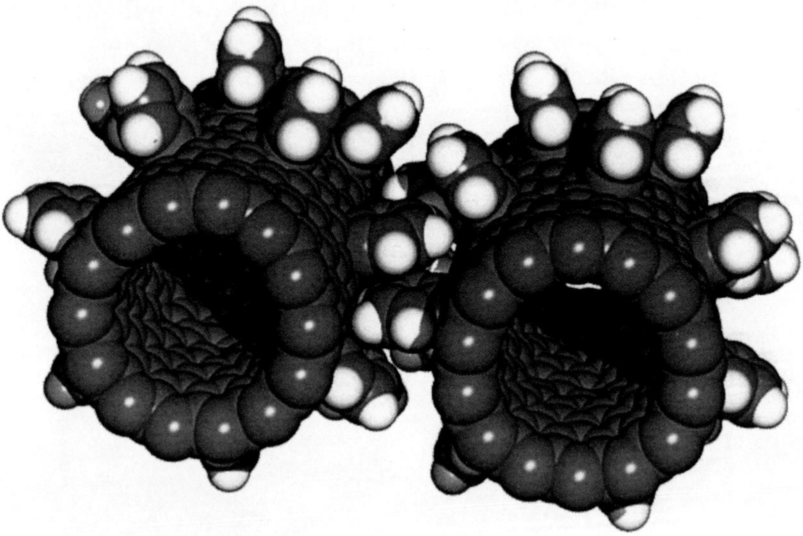

Figure: *Computer simulation of "nanogears" made of fullerene molecules. It is hoped that advances in nanoscience will lead to machines working on the molecular scale.*

Research in condensed matter physics has given rise to several device applications, such as the development of the semiconductor transistor, and laser technology. Several phenomena studied in the context of nanotechnology come under the purview of condensed matter physics. Techniques such as scanning-tunneling microscopy can be used to control processes at the nanometre scale, and have given rise to the study of nanofabrication. Several condensed matter systems are being studied with potential applications to quantum computation, including experimental systems like quantum dots, SQUIDs, and theoretical models like the toric code and the quantum dimer model. Condensed matter systems can be tuned to provide the conditions of coherence and phase-sensitivity that are essential ingredients for quantum information storage. Spintronics is a new area of technology that can be used for information processing and transmission, and is based on spin, rather than electron transport.

Crystallography

Crystallography is the science that examines the arrangement of atoms in solids. The word "crystallography" derives from the Greek words *crystallon* "cold drop, frozen drop," with its meaning extending to all solids with some degree of transparency, and *grapho* "I write." In July 2012, the United Nations recognised the importance of the science of crystallography by proclaiming that 2014 would be the International Year of Crystallography. X-ray crystallography is used to determine the structure of large biomolecules, such as proteins.

Before the development of X-ray diffraction crystallography, the study of crystals was based on their geometry. This involves measuring the angles of crystal faces relative to theoretical reference axes (crystallographic axes), and establishing the symmetry of the crystal in question. The former is carried out using a goniometre. The position in 3D space of each crystal face is plotted on a stereographic net, e.g. Wulff net or Lambert net. In fact, the pole to each face is plotted on the net. Each point is labelled with its Miller index. The final plot allows the symmetry of the crystal to be established.

Crystallographic methods now depend on the analysis of the diffraction patterns of a sample targeted by a beam of some type. Although X-rays are most commonly used, the beam is not always electromagnetic radiation. For some purposes electrons or neutrons are used. This is facilitated by the wave properties of the particles. Crystallographers often explicitly state the type of illumination used when referring to a method, as with the terms *X-ray diffraction, neutron diffraction* and *electron diffraction.*

These three types of radiation interact with the specimen in different ways. X-rays interact with the spatial distribution of the valence electrons, while electrons are charged particles and therefore feel the total charge distribution of both the atomic nuclei and the surrounding electrons. Neutrons are scattered by the atomic nuclei through the strong nuclear forces, but in addition, the magnetic moment of neutrons is non-zero. They are therefore also scattered by magnetic fields. When neutrons are scattered from hydrogen-containing materials, they produce diffraction patterns with high noise levels. However, the material can sometimes be treated to substitute deuterium for hydrogen. Because of these different forms of interaction, the three types of radiation are suitable for different crystallographic studies.

Generally, an image of a small object is made using a lens to focus the illuminating radiation, as is done with the rays of the visible spectrum in light microscopy. However, the wavelength of visible light (about 4000 to 7000 ångström) is three orders of magnitude longer than the length of typical atomic bonds and atoms themselves (about 1 to 2 Å). Therefore, obtaining information about the spatial arrangement of atoms requires the use of radiation with shorter wavelengths, such as X-ray or neutron beams.

Employing shorter wavelengths implied abandoning microscopy and true imaging, however, because there exists no material from which a lens capable of focusing this type of radiation can be created. (That said, scientists have had some success focusing X-rays with microscopic Fresnel zone plates made from gold, and by critical-angle reflection inside long tapered capillaries.)

Diffracted X-ray or neutrons beams cannot be focused to produce images, so the sample structure must be reconstructed from the diffraction pattern. Sharp features in the diffraction pattern arise from periodic, repeating structure in the sample, which are often very strong due to coherent reflection of many photons from many regularly spaced instances of similar structure, while non-periodic components of the structure result in diffuse (and usually weak) diffraction features. Said more simply, areas with a higher density and repetition of atom order tend to reflect more light toward one point in space when compared to those areas with fewer atoms and less repetition.

Because of their highly ordered and repetitive structure, crystals give diffraction patterns of sharp Bragg reflection spots, and are ideal for analyzing the structure of solids.

Notation

- Coordinates in *square brackets* such as [100] denote a direction vector (in real space).
- Coordinates in *angle brackets* or *chevrons* such as <100> denote a *family* of directions which are related by symmetry operations. In the cubic crystal system for example, <100> would mean [100], [010], [001] or the negative of any of those directions.
- Miller indices in *parentheses* such as (100) denote a plane of the crystal structure, and regular repetitions of that plane with a particular spacing. In the cubic system, the normal to the (hkl) plane is the direction [hkl], but in lower-symmetry cases, the normal to (hkl) is not parallel to [hkl].

- Indices in *curly brackets* or *braces* such as {100} denote a family of planes and their normals which are equivalent in cubic materials due to symmetry operations, much the way angle brackets denote a family of directions. In non-cubic materials, <hkl> is not necessarily perpendicular to {hkl}.

Technique

Some materials studied using crystallography, proteins for example, do not occur naturally as crystals. Typically, such molecules are placed in solution and allowed to crystallize over days, weeks, or months through vapor diffusion.

A drop of solution containing the molecule, buffer, and precipitants is sealed in a container with a reservoir containing a hygroscopic solution. Water in the drop diffuses to the reservoir, slowly increasing the concentration and allowing a crystal to form. If the concentration were to rise more quickly, the molecule would simply precipitate out of solution, resulting in disorderly granules rather than an orderly and hence usable crystal.

Once a crystal is obtained, data can be collected using a beam of radiation. Although many universities that engage in crystallographic research have their own X-ray producing equipment, synchrotrons are often used as X-ray sources, because of the purer and more complete patterns such sources can generate.

Synchrotron sources also have a much higher intensity of X-ray beams, so data collection takes a fraction of the time normally necessary at weaker sources. Complementary neutron crystallography techniques are used to enhance hydrogen atoms positions. Such techniques are available in Neutron facilities.

Producing an image from a diffraction pattern requires sophisticated mathematics and often an iterative process of modelling and refinement. In this process, the mathematically predicted diffraction patterns of an hypothesized or "model" structure are compared to the actual pattern generated by the crystalline sample. Ideally, researchers make several initial guesses, which through refinement all converge on the same answer. Models are refined until their predicted patterns match to as great a degree as can be achieved without radical revision of the model. This is a painstaking process, made much easier today by computers.

The mathematical methods for the analysis of diffraction data only apply to *patterns,* which in turn result only when waves diffract from orderly arrays. Hence crystallography applies for the most part only to crystals, or to molecules which can be coaxed to crystallize for the sake of measurement.

In spite of this, a certain amount of molecular information can be deduced from the patterns that are generated by fibres and powders, which while not as perfect as a solid crystal, may exhibit a degree of order. This level of order can be sufficient to deduce the structure of simple molecules, or to determine the coarse features of more complicated molecules. For example, the double-helical structure of DNA was deduced from an X-ray diffraction pattern that had been generated by a fibrous sample.

Crystallography in Materials Engineering

Crystallography is a tool that is often employed by materials scientists. In single crystals, the effects of the crystalline arrangement of atoms is often easy to see macroscopically, because the natural shapes of crystals reflect the atomic structure. In addition, physical properties are often controlled by crystalline defects. The understanding of crystal structures is an important prerequisite for understanding crystallographic defects. Mostly, materials do not occur in a single crystalline, but poly-crystalline form, such that the powder diffraction method plays a most important role in structural determination.

A number of other physical properties are linked to crystallography. For example, the minerals in clay form small, flat, platelike structures. Clay can be easily deformed because the platelike particles can slip along each other in the plane of the plates, yet remain strongly connected in the direction perpendicular to the plates. Such mechanisms can be studied by crystallographic texture measurements.

In another example, iron transforms from a body-centred cubic (bcc) structure to a face-centred cubic (fcc) structure called austenite when it is heated. The fcc structure is a close-packed structure, and the bcc structure is not, which explains why the volume of the iron decreases when this transformation occurs.

Crystallography is useful in phase identification. When performing any process on a material, it may be desired to find out what compounds and what phases are present in the material. Each phase has a characteristic arrangement of atoms. Techniques like X-ray or neutron diffraction can be used to identify which patterns are present in the material, and thus which compounds are present.

Crystallography covers the enumeration of the symmetry patterns which can be formed by atoms in a crystal and for this reason has a relation to group theory and geometry.

Crystallographic Database

A crystallographic database is a database specifically designed to store information about the structure of molecules and crystals. Crystals are solids having, in all three dimensions of space, a regularly repeating arrangement of atoms, ions, or molecules. They are characterized by symmetry, morphology, and directionally dependent physical properties. A crystal structure describes the arrangement of atoms, ions, or molecules in a crystal. (Molecules need to crystallize into solids so that their regularly repeating arrangements can be taken advantage of in X-ray, neutron, and electron diffraction based crystallography.)

Crystal structures of crystalline material are typically determined from X-ray or neutron single-crystal diffraction data and stored in crystal structure databases. They are routinely identified by comparing reflection intensities and lattice spacings from X-ray powder diffraction data with entries in powder-diffraction fingerprinting databases.

Crystal structures of nanometre sized crystalline samples can be determined via structure factor amplitude information from single-crystal electron diffraction data or structure factor amplitude and phase angle information from Fourier transforms of HRTEM images of crystallites. They are stored in crystal structure databases specialising in nanocrystals and can be identified by comparing zone axis subsets in lattice-fringe fingerprint plots with entries in a lattice-fringe fingerprinting database.

Crystallographic databases differ in access and usage rights and offer varying degrees of search and analysis capacity. Many provide structure visualization capabilities. They can be browser based or installed locally. Newer versions are built on the relational database model and support the Crystallographic Information File (CIF) as a universal data exchange format.

Crystallographic data are primarily extracted from published scientific articles and supplementary material. Newer versions of crystallographic databases are built on the relational database model, which enables efficient cross-referencing of tables. Cross-referencing serves to derive additional data or enhance the search capacity of the database.

Data exchange among crystallographic databases, structure visualization software, and structure refinement programs has been facilitated by the emergence of the Crystallographic Information File (CIF) format. The CIF format is the standard file format for the exchange and archiving of crystallographic data. It was adopted by the International

Union of Crystallography (IUCr), who also provides full specifications of the format. It is supported by all major crystallographic databases.

The increasing automation of the crystal structure determination process has resulted in ever higher publishing rates of new crystal structures and, consequentially, new publishing models. Minimalistic articles contain only crystal structure tables, structure images, and, possibly, abstract-like structure description. They tend to be published in author-financed or subsidized open-access journals. *Acta Crystallographica Section E* and *Zeitschrift für Kristallographie* belong in this category. More elaborate contributions may go to traditional subscriber-financed journals. Hybrid journals, on the other hand, embed individual author-financed open-access articles among subscriber-financed ones. Publishers may also make scientific articles available online, as Portable Document Format (PDF) files.

Crystal structure data in CIF format are linked to scientific articles as supplementary material. CIFs may be accessible directly from the publisher's website, crystallographic databases, or both. In recent years, many publishers of crystallographic journals have come to interpret CIFs as formatted versions of open data, i.e. representing non-copyrightable facts, and therefore tend to make them freely available online, independent of the accessibility status of linked scientific articles.

Trends

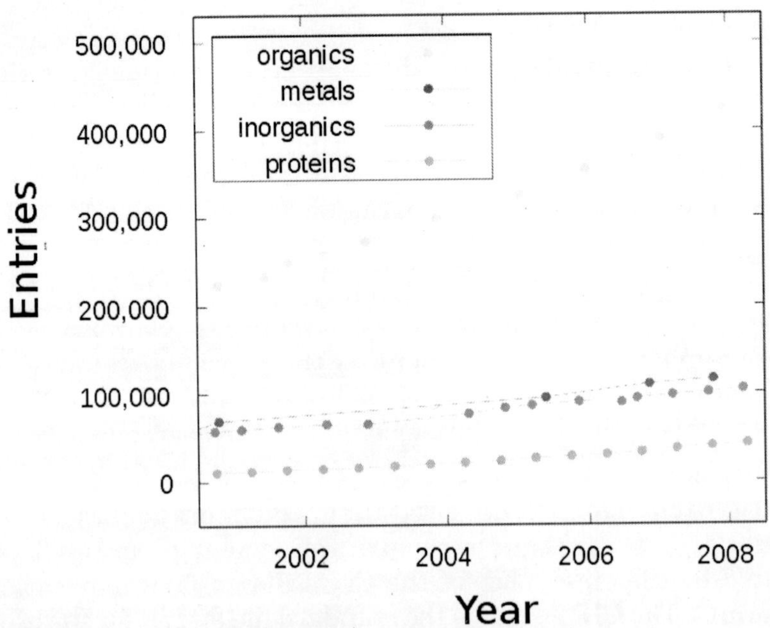

Figure: *Trends of crystal structures in databases over the last decade.*

As of 2008, more than 700,000 crystal structures had been published and stored in crystal structure databases. The publishing rate has reached more than 50,000 crystal structures per year. These numbers refer to published and republished crystal structures from experimental data. Crystal structures are republished owing to corrections for symmetry errors, improvements of lattice and atomic parameters, and differences in diffraction technique or experimental conditions. As of 2014, there are about 1,000,000 molecule and crystal structures known and published, probably a third of them in open access.

Crystal structures are typically categorized as minerals, metals-alloys, inorganics, organics, nucleic acids, and biological macromolecules. Individual crystal structure databases cater for users in specific chemical, molecular-biological, or related disciplines by covering super- or subsets of these categories. Minerals are a subset of mostly inorganic compounds. The category 'metals-alloys' covers metals, alloys, and intermetallics. Metals-alloys and inorganics can be merged into 'non-organics'. Organic compounds and biological macromolecules are separated according to molecular size. Organic salts, organometallics, and metalloproteins tend to be attributed to organics or biological macromolecules, respectively. Nucleic acids are a subset of biological macromolecules.

Comprehensiveness can refer to the number of entries in a database. On those terms, a crystal structure database can be regarded as comprehensive, if it contains a collection of all (re-)published crystal structures in the category of interest and is updated frequently. Searching for structures in such a database can replace more time-consuming scanning of the open literature. Access to crystal structure databases differs widely. It can be divided into reading and writing access. Reading access rights (search, download) affect the number and range of users. Restricted reading access is often coupled with restricted usage rights. Writing access rights (upload, edit, delete), on the other hand, determine the number and range of contributors to the database. Restricted writing access is often coupled with high data integrity.

In terms of user numbers and daily access rates, comprehensive and thoroughly vetted open-access crystal structure databases naturally surpass comparable databases with more restricted access and usage rights. Independent of comprehensiveness, open-access crystal structure databases have spawned open-source software projects, such as search-analysis tools, visualization software, and derivative databases. Scientific progress has been slowed down by restricting access or usage rights as well as limiting comprehensiveness or data integrity. Restricted access or usage rights are commonly associated with commercial crystal structure databases. Lack of comprehensiveness or data integrity, on the other hand, are associated with some of the open-access crystal structure databases other than the Crystallography Open Database (COD), and is "macromolecular open-access counterpart", the world wide Protein Database. Apart from that, several crystal structure databases are freely available for primarily educational purposes, in particular mineralogical databases and educational offshoots of the COD.

Crystallographic databases can specialise in crystal structures, crystal phase identification, crystallization, crystal morphology, or various physical properties. More integrative databases combine several categories of compounds or specialisations. Structures of incommensurate phases, nanocrystals, thin films on substrates, and predicted crystal structures are collected in tailored special structure databases.

Search

Search capacities of crystallographic databases differ widely. Basic functionality comprises search by keywords, physical properties, and chemical elements. Of particular importance is search by compound name and lattice parameters. Very useful are search options that allow the use of wildcard characters and logical connectives in search strings. If supported, the scope of the search can be constrained by the exclusion of certain chemical elements.

More sophisticated algorithms depend on the material type covered. Organic compounds might be searched for on the basis of certain molecular fragments. Inorganic compounds, on the other hand, might be of interest with regard to a certain type of coordination geometry. More advanced algorithms deal with conformation analysis (organics), supramolecular chemistry (organics), interpolyhedral connectivity ('non-organics') and higher-order molecular structures (biological macromolecules). Search algorithms used for a more complex analysis of physical properties, e.g. phase transitions or structure-property relationships, might apply group-theoretical concepts.

Modern versions of crystallographic databases are based on the relational database model. Communication with the database usually happens via a dialect of the Structured Query Language (SQL). Web-based databases typically process the search algorithm on the server interpreting supported scripting elements, while desktop-based databases run locally installed and usually precompiled search engines.

Crystal Phase Identification

Crystalline material may be divided into single crystals, twin crystals, polycrystals, and crystal powder. In a single crystal, the arrangement of atoms, ions, or molecules is defined by a single crystal structure in one orientation. Twin crystals, on the other hand, consist of single-crystalline twin domains, which are aligned by twin laws and separated by domain walls.

Polycrystals are made of a large number of small single crystals, or crystallites, held together by thin layers of amorphous solid. Crystal powder is obtained by grinding crystals, resulting in powder particles, made up of one or more crystallites. Both polycrystals and crystal powder consist of many crystallites with varying orientation.

Crystal phases are defined as regions with the same crystal structure, irrespective of orientation or twinning. Single and twinned crystalline specimens therefore constitute individual crystal phases. Polycrystalline or crystal powder samples may consist of more than one crystal phase. Such a phase comprises all the crystallites in the sample with the same crystal structure. Crystal phases can be identified by successfully matching suitable crystallographic parameters with their counterparts in database entries. Prior knowledge of the chemical composition of the crystal phase can be used to reduce the number of database entries to a small selection of candidate structures and thus simplify the crystal phase identification process considerably.

Powder Diffraction Fingerprinting (1D)

Applying standard diffraction techniques to crystal powders or polycrystals is tantamount to collapsing the 3D reciprocal space, as obtained via single-crystal diffraction, onto a 1D axis. The resulting partial-to-total overlap of symmetry-independent reflections renders the structure determination process more difficult, if not impossible. Powder diffraction data can be plotted as diffracted intensity (I) versus reciprocal lattice spacing ($1/d$). Reflection positions and intensities of known crystal phases, mostly from X-ray diffraction data, are stored, as d-I data pairs, in the Powder Diffraction File (PDF) database. The list of d-I data pairs is highly characteristic of a crystal phase and, thus, suitable for the identification, also called 'fingerprinting', of crystal phases.

Search-match algorithms compare selected test reflections of an unknown crystal phase with entries in the database. Intensity-driven algorithms utilise the three most intense lines (so-called 'Hanawalt search'), while d-spacing-driven algorithms are based on the eight to ten largest d-spacings (so-called 'Fink search').

X-ray powder diffraction fingerprinting has become the standard tool for the identification of single or multiple crystal phases and is widely used in such fields as metallurgy, mineralogy, forensic science, archeology, condensed matter physics, and the biological and pharmaceutical sciences.

Lattice-Fringe Fingerprinting (2D)

Powder diffraction patterns of very small single crystals, or crystallites, are subject to size-dependent peak broadening, which, below a certain size, renders powder diffraction fingerprinting useless. In this case, peak resolution is only possible in 3D reciprocal space, i.e. by applying single-crystal electron diffraction techniques.

High-Resolution Transmission Electron Microscopy (HRTEM) provides images and diffraction patterns of nanometre sized crystallites. Fourier transforms of HRTEM images and electron diffraction patterns both supply information about the projected reciprocal lattice geometry for a certain crystal orientation, where the projection axis coincides with the optical axis of the microscope.

Projected lattice geometries can be represented by so-called 'lattice-fringe fingerprint plots' (LFFPs), also called angular covariance plots. The horizontal axis of such a plot is given in reciprocal lattice length and is limited by the point resolution of the microscope. The vertical axis is defined as acute angle between Fourier transformed lattice fringes or electron diffraction spots. A 2D data point is defined by the length of a reciprocal lattice vector and its (acute) angle with another reciprocal lattice vector. Sets of 2D data points that obey Weiss's zone law are subsets of the entirety of data points in an LFFP. A suitable search-match algorithm using LFFPs, therefore, tries to find matching zone axis subsets in the database. It is, essentially, a variant of a lattice matching algorithm.

The performance of search-match procedures utilising LFFPs, also called 'lattice-fringe fingerprinting', can be sped up by precalculating and storing full LFFPs of all entries, assuming either kinematic or dynamic scattering and a given point resolution of the microscope. The number of possible entries can be narrowed down on the basis of chemical compound information.

In the case of electron diffraction patterns, structure factor amplitudes can be used, in a later step, to further discern among a selection of candidate structures (so-called 'structure factor fingerprinting'). Structure factor amplitudes from electron diffraction data are far less reliable than their counterparts from X-ray single-crystal and powder diffraction data. Existing precession electron diffraction techniques greatly improve the quality of structure factor amplitudes, increase their number and, thus, make structure factor amplitude information much more useful for the fingerprinting process.

Fourier transforms of HRTEM images, on the other hand, supply information not only about the projected reciprocal lattice geometry and structure factor amplitudes, but also structure factor phase angles. After crystallographic image processing, structure factor phase angles are far more reliable than structure factor amplitudes. Further discernment of candidate structures is then mainly based on structure factor phase angles and, to a lesser extent, structure factor amplitudes (so-called 'structure factor fingerprinting').

Morphological Fingerprinting (3D)

The Generalized Steno Law states that the interfacial angles between identical faces of any single crystal of the same material are, by nature, restricted to the same value. This offers the opportunity to fingerprint crystalline materials on the basis of optical goniometry, which is also known as crystallometry. In order to employ this technique successfully, one must consider the observed point group symmetry of the measured faces and creatively apply the rule that "crystal morphologies are often combinations of simple (i.e. low multiplicity) forms where the individual faces have the lowest possible Miller indices for any given zone axis". This shall ensure that the correct indexing of the crystal faces is obtained for any single crystal.

It is in many cases possible to derive the ratios of the crystal axes for crystals with low symmetry from optical goniometry with high accuracy and precision and to identify a crystalline material on their basis alone employing databases such as 'Crystal Data'. Provided that the crystal faces have been correctly indexed and the interfacial angles were measured to better than a few fractions of a tenth of a degree, a crystalline material can be identified quite unambiguously on the basis of angle comparisons to two rather comprehensive databases: the 'Bestimmungstabellen für Kristalle (Îïπåäåëθòåëü Êπθñòàëëîâ)' and the 'Barker Index of Crystals'.

Since Steno's Law can be further generalized for a single crystal of any material to include the angles between either all identically indexed net planes (i.e. vectors of the reciprocal lattice, also known as 'potential reflections in diffraction experiments') or all identically indexed lattice directions (i.e. vectors of the direct lattice, also known as zone axes), opportunities exist for morphological fingerprinting of nanocrystals in the transmission electron microscope (TEM) by means of transmission electron goniometry.

The specimen goniometre of a TEM is thereby employed analogously to the goniometre head of an optical goniometre. The optical axis of the TEM is then analogous to the reference direction of an optical goniometre. While in optical goniometry net-plane normals (reciprocal lattice vectors) need to be successively aligned parallel to the reference direction of an optical goniometre in order to derive measurements of interfacial angles, the corresponding alignment needs to be done for zone axes (direct lattice vector) in transmission electron goniometry. (Note that such alignments are by their nature quite trivial for nanocrystals in a TEM after the microscope has been aligned by standard procedures.)

Since transmission electron goniometry is based on Bragg's Law for the transmission (Laue) case (diffraction of electron waves), interzonal angles (i.e. angles between lattice directions) can be measured by a procedure that is analogous to the measurement of interfacial angles in an optical goniometre on the basis of Snell's Law, i.e. the reflection of light. The complements to interfacial angles of external crystal faces can, on the other hand, be directly measured from a zone-axis diffraction pattern or from the Fourier transform of a high resolution TEM image that shows crossed lattice fringes.

Lattice Matching (3D)

Lattice parameters of unknown crystal phases can be obtained from X-ray, neutron, or electron diffraction data. Single-crystal diffraction experiments supply orientation matrices, from which lattice parameters can be deduced. Alternatively, lattice parameters can be obtained from powder or polycrystal diffraction data via profile fitting without structural model (so-called 'Le Bail method').

Arbitrarily defined unit cells can be transformed to a standard setting and, from there, further reduced to a primitive smallest cell. Sophisticated algorithms compare such reduced cells with corresponding database entries. More powerful algorithms also consider derivative super- and subcells. The lattice-matching process can be further sped up by precalculating and storing reduced cells for all entries. The algorithm searches for matches within a certain range of the lattice parameters. More accurate lattice parameters allow a narrower range and, thus, a better match.

Lattice matching is useful in identifying crystal phases in the early stages of single-crystal diffraction experiments and, thus, avoiding unnecessary full data collection and structure determination procedures for already known crystal structures. The method is particularly important for single-crystalline samples that need to be preserved. If, on the other hand, some or all of the crystalline sample material can be ground, powder diffraction fingerprinting is usually the better option for crystal phase identification, provided that the peak resolution is good enough. However, lattice matching algorithms are still better at treating derivative super- and subcells.

Visualization

Newer versions of crystal structure databases integrate the visualization of crystal and molecular structures. Specialised or integrative crystallographic databases may provide morphology or tensor visualization output.

Crystal Structures

The crystal structure describes the three-dimensional periodic arrangement of atoms, ions, or molecules in a crystal. The unit cell represents the simplest repeating unit of the crystal structure. It is a parallelepiped containing a certain spatial arrangement of atoms, ions, molecules, or molecular fragments. From the unit cell the crystal structure can be fully reconstructed via translations.

The visualization of a crystal structure can be reduced to the arrangement of atoms, ions, or molecules in the unit cell, with or without cell outlines. Structure elements extending beyond single unit cells, such as isolated molecular or polyhedral units as well as chain, net, or framework structures, can often be better understood by extending the structure representation into adjacent cells.

The space group of a crystal is a mathematical description of the symmetry inherent in the structure. The motif of the crystal structure is given by the asymmetric unit, a minimal subset of the unit cell contents. The unit cell contents can be fully reconstructed via the symmetry operations of the space group on the asymmetric unit. Visualization interfaces usually allow for switching between asymmetric unit and full structure representations.

Bonds between atoms or ions can be identified by characteristic short distances between them. They can be classified as covalent, ionic, hydrogen, or other bonds including hybrid forms. Bond angles can be deduced from the bond vectors in groups of atoms or ions. Bond distances and angles can be made available to the user in tabular form or interactively, by selecting pairs or groups of atoms or ions. In ball-and-stick models of crystal structures, balls represent atoms and sticks represent bonds. Since organic chemists are particularly interested in molecular structures, it might be useful to be able to single out individual molecular units interactively from the drawing. Organic molecular units need to be given both as 2D structural formulae and full 3D molecular structures. Molecules on special-symmetry positions need to be reconstructed from the asymmetric unit. Protein crystallographers are interested in molecular structures of biological macromolecules, so that provisions need to be made to be able to represent molecular subunits as helices, sheets, or coils, respectively.

Crystal structure visualization can be integrated into a crystallographic database. Alternatively, the crystal structure data are exchanged between the database and the visualization software, preferably using the CIF format. Web-based crystallographic databases can integrate crystal structure visualization capability. Depending on the complexity of the structure, lighting, and 3D effects, crystal structure visualization can require a significant amount of processing power, which is why the actual visualization is typically run on the client.

Currently, web-integrated crystal structure visualization is based on Java applets from open-source projects such as Jmol. Web-integrated crystal structure visualization is tailored for examining crystal structures in web browsers, often supporting wide colour spectra (up to 32 bit) and window size adaptation. However, web-generated crystal structure images are not always suitable for publishing due to issues such as resolution depth, colour choice, grayscale contrast, or labelling (positioning, font type, font size).

Morphology and Physical Properties

Mineralogists, in particular, are interested in morphological appearances of individual crystals, as defined by the actually formed crystal faces (tracht) and their relative sizes (habit). More advanced visualization capabilities allow for displaying surface characteristics, imperfections inside the crystal, lighting (reflection, shadow, and translucency), and 3D effects (interactive rotatability, perspective, and stereo viewing). Crystal physicists, in particular, are interested in anisotropic physical properties of crystals. The directional dependence of a crystal's physical property is described by a 3D tensor and depends on the orientation of the crystal. Tensor shapes are more palpable by adding lighting effects (reflection and shadow). 2D sections of interest are selected for display by rotating the tensor interactively around one or more axes.

Crystal morphology or physical property data can be stored in specialised databases or added to more comprehensive crystal structure databases. The Crystal Morphology Database (CMD) is an example for a web-based crystal morphology database with integrated visualization capabilities.

Electron Diffraction

Electron diffraction refers to the wave nature of electrons. However, from a technical or practical point of view, it may be regarded as a technique used to study matter by firing electrons at a sample and observing the resulting interference pattern.

This phenomenon is commonly known as the wave–particle duality, which states that the *behavior* of a particle of matter (in this case the incident electron) can be described by a wave. For this reason, an electron can be regarded as a wave much like sound or water waves. This technique is similar to X-ray and neutron diffraction.

Electron diffraction is most frequently used in solid state physics and chemistry to study the crystal structure of solids. Experiments are usually performed in a transmission electron microscope (TEM), or a scanning electron microscope (SEM) as electron backscatter diffraction. In these instruments, electrons are accelerated by an electrostatic potential in order to gain the desired energy and determine their wavelength before they interact with the sample to be studied.

The periodic structure of a crystalline solid acts as a diffraction grating, scattering the electrons in a predictable manner. Working back from the observed diffraction pattern, it may be possible to deduce the structure of the crystal producing the diffraction pattern. However, the technique is limited by the phase problem. Apart from the study of crystals i.e. electron crystallography, electron diffraction is also a useful technique to study the short range order of amorphous solids, and the geometry of gaseous molecules.

Theory

Electron Interaction with Matter: Unlike other types of radiation used in diffraction studies of materials, such as X-rays and neutrons, electrons are charged particles and interact with matter through the Coulomb forces. This means that the incident electrons feel the influence of both the positively charged atomic nuclei and the surrounding electrons.

In comparison, X-rays interact with the spatial distribution of the valence electrons, while neutrons are scattered by the atomic nuclei through the strong nuclear forces.

In addition, the magnetic moment of neutrons is non-zero, and they are therefore also scattered by magnetic fields. Because of these different forms of interaction, the three types of radiation are suitable for different studies.

Intensity of Diffracted Beams

In the kinematical approximation for electron diffraction, the intensity of a diffracted beam is given by:

$$I_g = |\psi_g|^2 \propto |F_g|^2.$$

Here ψ_g is the wavefunction of the diffracted beam and F_g is the so-called structure factor which is given by:

$$F_g = \sum_i f_i e^{-2\pi i g \cdot r_i}$$

where g is the scattering vector of the diffracted beam, r_i is the position of an atom i in the unit cell, and f_i is the scattering power of the atom, also called the atomic form factor. The sum is over all atoms in the unit cell.

The structure factor describes the way in which an incident beam of electrons is scattered by the atoms of a crystal unit cell, taking into account the different scattering power of the elements through the factor f_i. Since the atoms are spatially distributed in the unit cell, there will be a difference in phase when considering the scattered amplitude from two atoms. This phase shift is taken into account by the exponential term in the equation.

The atomic form factor, or scattering power, of an element depends on the type of radiation considered. Because electrons interact with matter though different processes than for example X-rays, the atomic form factors for the two cases are not the same.

Wavelength of Electrons

The wavelength of an electron is given by the de Broglie equation

$$\lambda = \frac{h}{p}.$$

Here h is Planck's constant and p the relativistic momentum of the electron. λ is called the de Broglie wavelength. The electrons are accelerated in an electric potential U to the desired velocity:

$$v = \sqrt{\frac{2eU}{m_0}}$$

m_0 is the mass of the electron, and is the elementary charge. The electron wavelength is then given by:

$$\lambda = \frac{h}{p} = \frac{h}{m_0 v} = \frac{h}{\sqrt{2m_0 eU}}.$$

However, in an electron microscope, the accelerating potential is usually several thousand volts causing the electron to travel at an appreciable fraction of the speed of light. An SEM may typically operate at an accelerating potential of 10,000 volts (10 kV) giving an electron velocity approximately 20% of the speed of light, while a typical TEM can operate at 200 kV raising the electron velocity to 70% the speed of light. We therefore need to take relativistic effects into account. The relativistic relation between energy and momentum is $E^2=p^2c^2+m_0^2c^4$ and it can be shown that,

$$p=\sqrt{2m_0\Delta E+\frac{\Delta E^2}{c^2}}=\sqrt{2m_0\Delta E}\sqrt{1+\frac{\Delta E}{2m_0c^2}}$$

where $\Delta E = E - E_0 = eU$. The relativistic formula for the wavelength is then modified to become,

$$\lambda=\frac{h}{\sqrt{2m_0eU}}\frac{1}{\sqrt{1+\frac{eU}{2m_0c^2}}}$$

c is the speed of light. We recognise the first term in this final expression as the non-relativistic expression derived above, while the last term is a relativistic correction factor. The wavelength of the electrons in a 10 kV SEM is then 12.2×10^{-12} m (12.2 pm) while in a 200 kV TEM the wavelength is 2.5 pm. In comparison the wavelength of X-rays usually used in X-ray diffraction is in the order of 100 pm (Cu kα: λ=154 pm).

Electron Diffraction in a TEM

Electron diffraction of solids is usually performed in a Transmission Electron Microscope (TEM) where the electrons pass through a thin film of the material to be studied. The resulting diffraction pattern is then observed on a fluorescent screen, recorded on photographic film, on imaging plates or using a CCD camera.

Benefits

As mentioned above, the wavelength of an electron accelerated in a TEM is much smaller than that of the radiation usually used for X-ray diffraction experiments. A consequence of this is that the radius of the Ewald sphere is much larger in electron diffraction experiments than in X-ray diffraction. This allows the diffraction experiment to reveal more of the two-dimensional distribution of reciprocal lattice points.

Furthermore, electron lenses allows the geometry of the diffraction experiment to be varied. The conceptually simplest geometry referred to as selected area electron diffraction (SAED) is that of a parallel beam of electrons incident on the specimen, with the specimen field selected using a sub-specimen image-plane aperture. However, by converging the electrons in a cone onto the specimen, one can in effect perform a diffraction experiment over several incident angles simultaneously. This technique is called Convergent Beam Electron Diffraction (CBED) and can reveal the full three-dimensional symmetry of the crystal.

In a TEM, a single crystal grain or particle may be selected for the diffraction experiments. This means that the diffraction experiments can be performed on single crystals of nanometre size, whereas other diffraction techniques would be limited to studying the diffraction from a multicrystalline or powder sample. Furthermore, electron diffraction in TEM can be combined with direct imaging of the sample, including high resolution imaging of the crystal lattice, and a range of other techniques. These include solving and refining crystal structures by electron crystallography, chemical analysis of the sample composition through energy-dispersive X-ray spectroscopy, investigations of electronic structure and bonding through electron energy loss spectroscopy, and studies of the mean inner potential through electron holography.

Practical Aspects

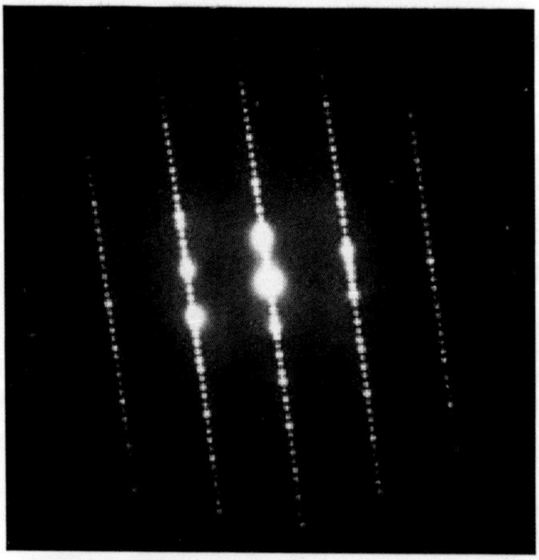

Figure: *Typical electron diffraction pattern obtained in a TEM with a parallel electron beam*

Figure depicted earlier, to the right is a simple sketch of the path of a parallel beam of electrons in a TEM from just above the sample and down the column to the fluorescent screen. As the electrons pass through the sample, they are scattered by the electrostatic potential set up by the constituent elements. After the electrons have left the sample they pass through the electromagnetic objective lens. This lens acts to collect all electrons scattered from one point of the sample in one point on the fluorescent screen, causing an image of the sample to be formed. We note that at the dashed line in the figure, electrons scattered in the same direction by the sample are collected into a single point. This is the back focal plane of the microscope, and is where the diffraction pattern is formed. By manipulating the magnetic lenses of the microscope, the diffraction pattern may be observed by projecting it onto the screen instead of the image. An example of what a diffraction pattern obtained in this way may look like is shown in figure above.

If the sample is tilted with respect to the incident electron beam, one can obtain diffraction patterns from several crystal orientations. In this way, the reciprocal lattice of the crystal can be mapped in three dimensions. By studying the systematic absence of diffraction spots the Bravais lattice and any screw axes and glide planes present in the crystal structure may be determined.

Limitations

Electron diffraction in TEM is subject to several important limitations. First, the sample to be studied must be electron transparent, meaning the sample thickness must be of the order of 100 nm or less. Careful and time consuming sample preparation may therefore be needed. Furthermore, many samples are vulnerable to radiation damage caused by the incident electrons.

The study of magnetic materials is complicated by the fact that electrons are deflected in magnetic fields by the Lorentz force. Although this phenomenon may be exploited to study the magnetic domains of materials by *Lorentz force microscopy*, it may make crystal structure determination virtually impossible. Furthermore, electron diffraction is often regarded as a *qualitative* technique suitable for symmetry determination, but too inaccurate for determination of lattice parameters and atomic positions. But there are also several examples where unknown crystal structures (both inorganic, organic and biological) have been solved by electron crystallography. Lattice parameters of high accuracy can in fact be obtained from electron diffraction, relative errors less than 0.1% have been demonstrated. However, the right experimental conditions may be difficult to obtain, and these procedures are often viewed as too time consuming and the data too difficult to interpret. X-ray or neutron diffraction are therefore often the preferred methods for determining lattice parameters and atomic positions.

However, the main limitation of electron diffraction in TEM remains the comparatively high level of user interaction needed. Whereas both the execution of powder X-ray (and neutron) diffraction experiments and the data analysis are highly automated and routinely performed, electron diffraction requires a much higher level of user input.

Electron Crystallography

Electron crystallography is a method to determine the arrangement of atoms in solids using a transmission electron microscope (TEM).

It can complement X-ray crystallography for studies of very small crystals (<0.1 micrometres), both inorganic, organic and proteins, such as membrane proteins, that cannot easily form the large 3-dimensional crystals required for that process.

Protein structures are usually determined from either 2-dimensional crystals (sheets or helices), polyhedrons such as viral capsids, or dispersed individual proteins. Electrons can be used in these situations, whereas X-rays cannot, because electrons interact more strongly with atoms than X-rays do.

Thus, X-rays will travel through a thin 2-dimensional crystal without diffracting significantly, whereas electrons can be used to form an image. Conversely, the strong interaction between electrons and proteins makes thick (e.g. 3-dimensional > 1 micrometre) crystals impervious to electrons, which only penetrate short distances.

One of the main difficulties in X-ray crystallography is determining phases in the diffraction pattern. Because of the complexity of X-ray lenses, it is difficult to form an image of the crystal being diffracted, and hence phase information is lost. Fortunately, electron microscopes contain electron lenses, and so the crystallographic structure factor phase information can be experimentally determined in electron crystallography. Aaron Klug was the first to realise that the phase information could be read out directly from the Fourier transform of an electron microscopy image that had been scanned into a computer, already in 1968. For this, and his studies on virus structures and transfer-RNA, Klug received the Nobel Prize for chemistry in 1982.

Radiation Damage

A common problem to X-ray crystallography and electron crystallography is radiation damage, by which especially organic molecules and proteins are damaged as they are being imaged, limiting the resolution that can be obtained. This is especially troublesome in the setting of electron crystallography, where that radiation damage is focused on far fewer atoms. One technique used to limit radiation damage is electron cryomicroscopy, in which the samples undergo cryofixation and imaging takes place at liquid nitrogen or even liquid helium temperatures. Because of this problem, X-ray crystallography has been much more successful in determining the structure of proteins that are especially vulnerable to radiation damage.

Protein Structures Determined by Electron Crystallography

The first electron crystallographic protein structure to achieve atomic resolution was bacteriorhodopsin, determined by Richard Henderson and coworkers at the Medical Research Council Laboratory of Molecular Biology in 1990.

However, already in 1975 Unwin and Henderson had determined the first membrane protein structure at intermediate resolution (7 Ångström), showing for the first time the internal structure of a membrane protein, with its alpha-helices standing perpendicular to the plane of the membrane.

Since then, several other high-resolution structures have been determined by electron crystallography, including the light-harvesting complex, the nicotinic acetylcholine receptor, and the bacterial flagellum.

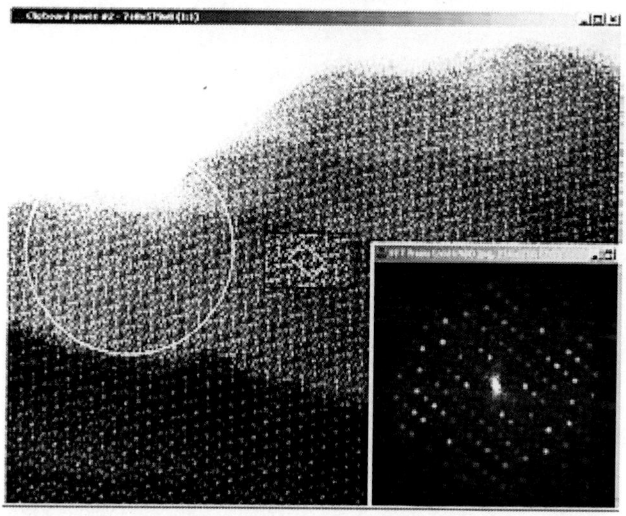

Figure: *Electron microscopy image of an inorganic tantalum oxide, with its Fourier transform, also called diffractogram, inset. Notice how the appearance changes from the upper thin region to the thicker lower region. The unit cell of this compound is about 15 by 25 Ångström. It is outlined at the centre of the figure, inside the result from image processing, where the symmetry has been taken into account. The black dots show clearly all the tantalum atoms. The diffraction extends to 6 orders along the 15 Å direction and 10 orders in the perpendicular direction. Thus the resolution of the EM image is 2.5 Å (15/6 or 25/10). This calculated Fourier transform contain both amplitudes (as seen) and phases (not displayed).*

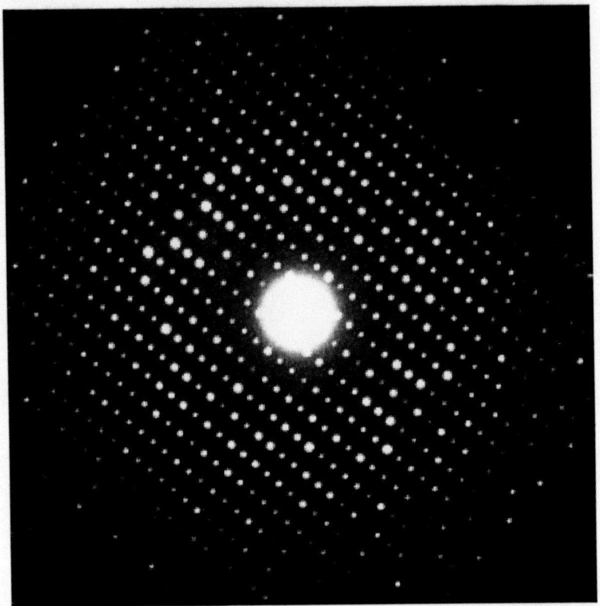

Figure: *Electron diffraction pattern of the same crystal of inorganic tantalum oxide shown above. Notice that there are many more diffraction spots here than in the diffractogram calculated from the EM image above. The diffraction extends to 12 orders along the 15 Å direction and 20 orders in the perpendicular direction. Thus the resolution of the ED pattern is 1.25 Å (15/12 or 25/20). ED patterns do not contain phase information, but the clear differences between intensities of the diffraction spots can be used in crystal structure determination.*

Application to Inorganic Materials

Electron crystallographic studies on inorganic crystals using high-resolution electron microscopy (HREM) images were first performed by Aaron Klug in 1978 and by Sven Hovmöller and coworkers in 1984. HREM images were used because they allow to select (by computer software) only the very thin regions close to the edge of the crystal for structure analysis. This is of crucial importance since in the thicker parts of the crystal the exit-wave function (which carries the information about the intensity and position of the projected atom columns) is no longer linearly related to the projected crystal structure.

Moreover not only do the HREM images change their appearance with increasing crystal thickness, they are also very sensitive to the chosen setting of the defocus Δf of the objective lens. To cope with this complexity Michael O'Keefe started in the early 1970s to develop image simulation software which allowed to understand an interpret the observed contrast changes in HREM images.

There was a serious disagreement in the field of electron microscopy of inorganic compounds; while some have claimed that "the phase information is present in EM images" others have the opposite view that "the phase information is lost in EM images". The reason for these opposite views is that the word "phase" has been used with different meanings in the two communities of physicists and crystallographers. The physicists are more concerned about the "electron wave phase" - the phase of a wave moving through the sample during exposure by the electrons. This wave has a wavelength of about 0.02-0.03 Ångström (depending on the accelerating voltage of the electron microscope). Its phase is related to the phase of the undiffracted direct electron beam. The crystallographers, on the other hand, mean the "crystallographic structure factor phase" when they simply say "phase".

This phase is the phase of standing waves of potential in the crystal (very similar to the electron density measured in X-ray crystallography). Each of these waves have their specific wavelength, called d-value for distance between so-called Bragg planes of low/high potential. These d-values range from the unit cell dimensions to the resolution limit of the electron microscope, i.e. typically from 10 or 20 Ångströms down to 1 or 2 Ångströms. Their phases are related to a fixed point in the crystal, defined in relation to the symmetry elements of that crystal. The crystallographic phases are a property of the crystal, so they exist also outside the electron microscope. The electron waves vanish if the microscope is switched off. In order to determine a crystal structure, it is necessary to know the crystallographic structure factors, but not to know the electron wave phases. A more detailed discussion how (crystallographic structure factor) phases link with the phases of the electron wave can be found in.

Just as with proteins, it has been possible to determine the atomic structures of inorganic crystals by electron crystallography. For simpler structure it is sufficient to use three perpendicular views, but for more complicated structures, also projections down ten or more different diagonals may be needed.

In addition to electron microscopy images, it is also possible to use electron diffraction (ED) patterns for crystal structure determination. The utmost care must be taken to record such ED patterns from the thinnest areas in order to keep most of the structure related intensity differences between the reflections (quasi-kinematical diffraction conditions). Just as with X-ray diffraction patterns, the important crystallographic structure factor phases are lost in electron diffraction patterns and must be uncovered by special crystallographic methods such as direct methods, maximum likelihood or (more recently) by the charge-flipping method. On the other hand, ED patterns of inorganic crystals have often a high resolution (= interplanar spacings with high Miller indices) much below 1 Ångström. This is comparable to the point resolution of the best electron microscopes. Under favourable conditions it is possible to use ED patterns from a single orientation to determine the complete crystal structure. Alternatively a hybrid approach can be used which uses HRTEM images for solving and intensities from ED for refining the crystal structure.

Recent progress for structure analysis by ED was made by introducing the Vincent-Midgley precession technique for recording electron diffraction patterns. The thereby obtained intensities are usually much closer to the kinematical intensities, so that even structures can be determined that are out of range when processing conventional (selected area) electron diffraction data.

Crystal structures determined via electron crystallography can be checked for their quality by using first-principles calculations within density functional theory (DFT). This approach was for the first time applied for the validation of several metal-rich structures which were only accessible by HRTEM and ED, respectively.

Recently, two very complicated zeolite structures have been determined by electron crystallography combined with X-ray powder diffraction. These are more complex than the most complex zeolite structures determined by X-ray crystallography.

4

Nanotechnology and Materials Science

Materials science, also commonly known as materials engineering or materials science and engineering, is an interdisciplinary field applying the properties of matter to various areas of science and engineering. This relatively new scientific field investigates the relationship between the structure of materials at atomic or molecular scales and their macroscopic properties. It incorporates elements of applied physics and chemistry, with significant media attention focused on Nano science and nanotechnology. In recent years, materials science is becoming more widely known as a specific field of science and engineering. It is an important part of forensic engineering (Forensic engineering is the investigation of materials, products, structures or components that fail or do not operate or function as intended, causing personal injury or damage to property.) and failure analysis, the latter being the key to understanding, for example, the cause of various aviation accidents. Many of the most pressing scientific problems that are faced today are due to the limitations of the materials that are available and, as a result, breakthroughs in this field are likely to have a significant impact on the future of technology.

The material of choice of a given era is often a defining point. Phrases such as Stone Age, Bronze Age, Iron Age, and Steel Age are great examples. Originally deriving from the manufacture of ceramics and its putative derivative metallurgy, materials science is one of the oldest forms of engineering and applied science. Modern materials science evolved directly from metallurgy, which itself evolved from mining and (likely) ceramics and the use of fire. A major breakthrough in the understanding of materials occurred in the late 19th century, when the American scientist Josiah Willard Gibbs demonstrated that the thermodynamic properties related to atomic structure in various phases are related to the physical properties of a material. Important elements of modern materials science are a product of the space race: the understanding and engineering of the metallic alloys, and silica and carbon materials, used in the construction of space vehicles enabling the exploration of space. Materials science has driven, and been driven by, the development of revolutionary technologies such as plastics, semiconductors, and biomaterials.

Before the 1960s (and in some cases decades after), many *materials science* departments were named *metallurgy* departments, from a 19th and early 20th century emphasis on metals. The field has since broadened to include every class of materials, including ceramics, polymers, semiconductors, magnetic materials, medical implant materials, biological materials and nanomaterials (materiomics).

Fundamentals

The basis of materials science involves relating the desired properties and relative performance of a material in a certain application to the structure of the atoms and phases in that material through characterization. The major determinants of the structure of a material and thus of its properties are its constituent chemical elements and the way in which it has been processed into its final form. These characteristics, taken together and related through the laws of thermodynamics, govern a material's microstructure, and thus its properties. The manufacture of a perfect crystal of a material is physically impossible. Instead materials scientists manipulate the defects in crystalline materials such as precipitates, grain boundaries (Hall–Petch relationship), interstitial atoms, vacancies or substitutional atoms, to create materials with the desired properties.

Not all materials have a regular crystal structure. Polymers display varying degrees of crystallinity, and many are completely non-crystalline. Glass as, some ceramics, and many natural materials are amorphous, not possessing any long-range order in their atomic arrangements. The study of polymers combines elements of chemical and statistical thermodynamics to give thermodynamic, as well as mechanical, descriptions of physical properties.

In addition to industrial interest, materials science has gradually developed into a field which provides tests for condensed matter or solid state theories. New physics emerge because of the diverse new material properties that need to be explained.

Classes of Materials

Materials science encompasses various classes of materials, each of which may constitute a separate field. There are several ways to classify materials. For instance by the type of bonding between the atoms. The traditional groups are ceramics, metals and polymers based on atomic structure and chemical composition. New materials have resulted in more classes. One way of classifying materials is:

- Biomaterials
- Carbon
- Ceramics
- Composite materials
- Glass
- Metals
- Nanomaterials
- Polymers
- Refractory
- Semiconductors
- Thin Films
- Functionally Graded Materials

Materials in Industry

Radical materials advances can drive the creation of new products or even new industries, but stable industries also employ materials scientists to make incremental improvements and troubleshoot issues with currently used materials. Industrial applications of materials science include materials design, cost-benefit tradeoffs in industrial production of materials, processing techniques (casting, rolling, welding, ion implantation, crystal growth, thin-film deposition, sintering, glassblowing, etc.), and analytical techniques (characterization techniques such as electron microscopy, x-ray diffraction, calorimetry, nuclear microscopy (HEFIB), Rutherford backscattering, neutron diffraction, small-angle X-ray scattering (SAXS), etc.).

Besides material characterization, the material scientist/engineer also deals with the extraction of materials and their conversion into useful forms. Thus ingot casting, foundry techniques, blast furnace extraction, and electrolytic extraction are all part of the required knowledge of a metallurgist/engineer. Often the presence, absence or variation of minute quantities of secondary elements and compounds in a bulk material will have a great impact on the final properties of the materials produced, for instance, steels are classified based on 1/10 and 1/100 weight percentages of the carbon and other alloying elements they contain. Thus, the extraction and purification techniques employed in the extraction of iron in the blast furnace will have an impact of the quality of steel that may be produced.

The overlap between physics and materials science has led to the offshoot field of *materials physics*, which is concerned with the physical properties of materials. The approach is generally more macroscopic and applied than in condensed matter physics.

Ceramics and Glasses

Figure: *Si_3N_4 ceramic bearing parts*

Another application of the material sciences is the structures of glass and ceramics, typically associated with the most brittle materials. Bonding in ceramics and glasses use covalent and ionic-covalent types with SiO_2 (silica or sand) as a fundamental building block. Ceramics are as soft as clay and as hard as stone and concrete. Usually, they are crystalline in form. Most glasses contain a metal oxide fused with silica. At high temperatures used to prepare glass, the material is a viscous liquid. The structure of glass forms into an amorphous state upon cooling. Windowpanes and eyeglasses are important examples. Fibres of glass are also available. Scratch resistant Corning Gorilla Glass is a well-known example of the application of materials science to drastically improve the properties of common components. Diamond and carbon in its graphite form are considered to be ceramics.

Engineering ceramics are known for their stiffness and stability under high temperatures, compression and electrical stress. Alumina, silicon carbide, and tungsten carbide are made from a fine powder of their constituents in a process of sintering with a binder. Hot pressing provides higher density material. Chemical vapor deposition can place a film of a ceramic on another material. Cermets are ceramic particles containing some metals. The wear resistance of tools is derived from cemented carbides with the metal phase of cobalt and nickel typically added to modify properties.

Composite Materials

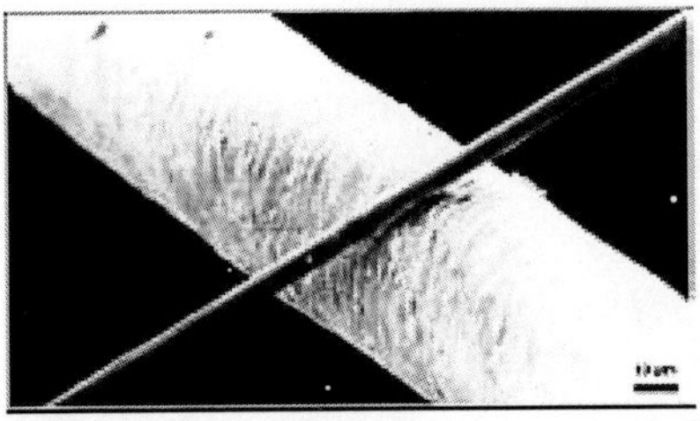

Figure: *A 6 μm diameter carbon filament (running from bottom left to top right) siting atop the much larger human hair.*

Filaments are commonly used for reinforcement in composite materials.

Another application of material science in industry is the making of composite materials. Composite materials are structured materials composed of two or more macroscopic phases. Applications range from structural elements such as steel-reinforced concrete, to the thermally insulative tiles which play a key and integral role in NASA's Space Shuttle thermal protection system which is used to protect the surface of the shuttle from the heat of re-entry into the Earth's atmosphere. One example is reinforced Carbon-Carbon (RCC), the light gray material which withstands re-entry temperatures up to 1510 °C (2750 °F) and protects the Space Shuttle's wing leading edges and nose cap. RCC is a laminated composite material made from graphite rayon cloth and impregnated with a phenolic resin. After curing at high temperature in an autoclave, the laminate is pyrolized to convert the resin to carbon, impregnated with furfural alcohol in a vacuum chamber, and cured/

pyrolized to convert the furfural alcohol to carbon. In order to provide oxidation resistance for reuse capability, the outer layers of the RCC are converted to silicon carbide.

Other examples can be seen in the "plastic" casings of television sets, cell-phones and so on. These plastic casings are usually a composite material made up of a thermoplastic matrix such as acrylonitrile-butadiene-styrene (ABS) in which calcium carbonate chalk, talc, glass fibres or carbon fibres have been added for added strength, bulk, or electrostatic dispersion. These additions may be referred to as reinforcing fibres, or dispersants, depending on their purpose.

Polymers

Polymers are also an important part of materials science. Polymers are the raw materials (the resins) used to make what we commonly call plastics. Plastics are really the final product, created after one or more polymers or additives have been added to a resin during processing, which is then shaped into a final form. Polymers which have been around, and which are in current widespread use, include polyethylene, polypropylene, PVC, polystyrene, nylons, polyesters, acrylics, polyurethanes, and polycarbonates. Plastics are generally classified as "commodity", "speciality" and "engineering" plastics.

PVC (polyvinyl-chloride) is widely used, inexpensive, and annual production quantities are large. It lends itself to an incredible array of applications, from artificial leather to electrical insulation and cabling, packaging and containers. Its fabrication and processing are simple and well-established. The versatility of PVC is due to the wide range of plasticisers and other additives that it accepts. The term "additives" in polymer science refers to the chemicals and compounds added to the polymer base to modify its material properties.

Polycarbonate would be normally considered an engineering plastic (other examples include PEEK, ABS). Engineering plastics are valued for their superior strengths and other special material properties. They are usually not used for disposable applications, unlike commodity plastics. Speciality plastics are materials with unique characteristics, such as ultra-high strength, electrical conductivity, electro-fluorescence, high thermal stability, etc.

The dividing lines between the various types of plastics is not based on material but rather on their properties and applications. For instance, polyethylene (PE) is a cheap, low friction polymer commonly used to make disposable shopping bags and trash bags, and is considered a commodity plastic, whereas medium-density polyethylene (MDPE) is used for underground gas and water pipes, and another variety called Ultra-high Molecular Weight Polyethylene UHMWPE is an engineering plastic which is used extensively as the glide rails for industrial equipment and the low-friction socket in implanted hip joints.

Metal Alloys

The study of metal alloys is a significant part of materials science. Of all the metallic alloys in use today, the alloys of iron (steel, stainless steel, cast iron, tool steel, alloy steels) make up the largest proportion both by quantity and commercial value. Iron alloyed with various proportions of carbon gives low, mid and high carbon steels. An iron carbon alloy is only considered steel if the carbon level is between 0.01% and 2.00%. For the steels, the hardness and tensile strength of the steel is related to the amount of carbon present, with increasing carbon levels also leading to lower ductility and toughness. Heat treatment processes such as quenching and tempering can significantly change these properties however. Cast Iron is defined as an iron–carbon alloy with more than 2.00% but less than 6.67% carbon. Stainless steel is defined as a regular steel alloy with greater than 10% by weight alloying content of Chromium. Nickel and Molybdenum are typically also found in stainless steels.

Other significant metallic alloys are those of aluminium, titanium, copper and magnesium. Copper alloys have been known for a long time (since the Bronze Age), while the alloys of thc other three metals have been relatively recently developed. Due to the chemical reactivity of these metals, the electrolytic extraction processes required were only developed relatively recently. The alloys of aluminium, titanium and magnesium are also known and valued for their high strength-to-weight ratios and, in the case of magnesium, their ability to provide electromagnetic shielding. These materials are ideal for situations where high strength-to-weight ratios are more important than bulk cost, such as in the aerospace industry and certain automotive engineering applications.

Sub-Disciplines of Materials Science

Below is a list of disciplines within or related to the materials science field. These range from biomaterials, to ceramics, to metals, to textile reinforced materials. Also note that these are linked to the respective main article.

- Biomaterials – materials that are derived from and/or used with life forms.
- Ceramography – the study of the microstructures of high-temperature materials and refractories, including structural

ceramics such as RCC, polycrystalline silicon carbide and transformation toughened ceramics

- Crystallography – the study of regular arrangement of atoms and ions in a solid, the defects associated with crystal structures such as grain boundaries and dislocations, and the characterization of these structures and their relation to physical properties.
- Electronic and magnetic materials – materials such as semiconductors used to create integrated circuits, storage media, sensors, and other devices.
- Forensic engineering – the study of how products fail, and the vital role of the materials of construction
- Forensic materials engineering – the study of material failure, and the light it sheds on how engineers specify materials in their product
- Glass science – any non-crystalline material including inorganic glasses, vitreous metals and non-oxide glasses.
- Materials characterization – such as diffraction with x-rays, electrons, or neutrons, and various forms of spectroscopy and chemical analysis such as Raman spectroscopy, energy-dispersive spectroscopy (EDS), chromatography, thermal analysis, electron microscope analysis, etc., in order to understand and define the properties of materials.
- Metallography - Metallography is the study of the physical structure and components of metals, typically using microscopy.
- Metallurgy – the study of metals and their alloys, including their extraction, microstructure and processing.
- Microtechnology – study of materials and processes and their interaction, allowing microfabrication of structures of micrometric dimensions, such as Microelectromechanical systems (MEMS).
- Nanotechnology – rigorously, the study of materials where the effects of quantum confinement, the Gibbs–Thomson effect, or any other effect only present at the nanoscale is the defining property of the material; but more commonly, it is the creation and study of materials whose defining structural properties are anywhere from less than a nanometre to one hundred nanometres in scale, such as molecularly engineered materials.
- Rheology – Some practitioners consider rheology a sub-field of materials science, because it can cover any material that flows.

However, modern rheology typically deals with non-Newtonian fluid dynamics, so it is often considered a sub-field of continuum mechanics.

- Surface science/catalysis – interactions and structures between solid-gas solid-liquid or solid-solid interfaces.
- Textile reinforced materials – materials in the form of ceramic or concrete are reinforced with a primarily woven or non-woven textile structure to impose high strength with comparatively more flexibility to withstand vibrations and sudden jerks.
- Tribology – the study of the wear of materials due to friction and other factors.

Methods, Processes, and Related Topics

Below are links to topics that explain methods, processes and related topics in order to enhance understanding of materials science.

- Alloying, corrosion, and thermal or mechanical processing, for a specialised treatment of metallurgical materials—with applications ranging from aerospace and industrial equipment to the civil industries
- Biomaterials, physiology, biomechanics, biochemistry, for a specialised understanding of how materials integrate into biological systems, e.g., through materiomics
- Crystallography, quantum chemistry or quantum physics, for the structure (symmetry and defects) and bonding in materials (e.g., ionic, metallic, covalent, and van der Waals bonding)
- Diffraction and wave mechanics, for the science behind characterization systems, e.g., X-ray diffraction (XRD) transmission electron microscopy (TEM)
- Electronic properties of materials, and solid-state physics, for the understanding of the electronic, thermal, magnetic, and optical properties of materials
- Mechanical behavior of materials, to understand the mechanical properties of materials, defects and their propagation, and their behavior under static, dynamic, and cyclic loads
- Phase transformation kinetics, for the kinetics of phase transformations (with particular emphasis on solid-solid phase transitions)
- Polymer properties, synthesis, and characterization, for a specialised understanding of how polymers behave, how they

are made, and how they are characterized; exciting applications of polymers include liquid crystal displays (LCDs, the displays found in most cell phones, cameras, and iPods), novel photovoltaic devices based on semiconductor polymers (which, unlike the traditional silicon solar panels, are flexible and cheap to manufacture, albeit with lower efficiency), and membranes for room-temperature fuel cells (as proton exchange membranes) and filtration systems in the environmental and biomedical fields

- Semiconductor materials and semiconductor devices, for a specialised understanding of the advanced processes used in industry (e.g. crystal growth techniques, thin-film deposition, ion implantation, photolithography), their properties, and their integration in electronic devices
- Solid-state physics is the study of rigid matter, or solids, through methods such as quantum mechanics, crystallography, electromagnetism, and metallurgy. It is the largest branch of condensed matter physics. Solid-state physics studies how the large-scale properties of solid materials result from their atomic-scale properties. Thus, solid-state physics forms the theoretical basis of materials science. It also has direct applications, for example in the technology of transistors and semiconductors.
- Thermodynamics, statistical mechanics, and physical chemistry, for phase equilibrium conditions, phase diagrams of materials systems (multi-phase, multi-component, reacting and non-reacting systems)

Building Material

Building material is any material which is used for construction purposes. Many naturally occurring substances, such as clay, rocks, sand, and wood, even twigs and leaves, have been used to construct buildings. Apart from naturally occurring materials, many man-made products are in use, some more and some less synthetic. The manufacture of building materials is an established industry in many countries and the use of these materials is typically segmented into specific speciality trades, such as carpentry, insulation, plumbing, and roofing work. They provide the make-up of habitats and structures including homes.

In history there are trends in building materials from being: natural to becoming more man-made and composite; biodegradable to imperishable; indigenous (local) to being transported globally; repairable to disposable; and chosen for increased levels of fire-safety. These trends tend to increase the *initial* and *long term* economic, ecological, energy, and social costs of building materials.

Economic Costs

The initial economic cost of building materials is the purchase price. This is often what governs decision making about what materials to use. Sometimes people take into consideration the energy savings or durability of the materials and see the value of paying a higher initial cost in return for a lower lifetime cost.

For example an asphalt shingle roof costs less than a metal roof to install, but the metal roof will last longer so the lifetime cost is less per year. Risks when considering lifetime cost of a material is if the building is damaged such as by fire or wind, or if the material is not as durable as advertised.

Ecological Costs

Pollution costs can be macro and micro. The macro, environmental pollution of extraction industries building materials rely on such as mining, petroleum, and logging produce environmental damage at there source and in transportation of the raw materials, manufacturing, transportation of the products, retailing, and installation. An example of the micro aspect of pollution is the off-gassing of the building materials in the building or indoor air pollution. Red List building materials are materials found to be harmful.

Also the carbon footprint, the total set of greenhouse gas emissions produced in the life of the material. A life-cycle analysis also includes the reuse, recycling, or disposal of construction waste. Two concepts in building which account for the ecological economics of building materials are green building and sustainable development.

Energy Costs

Initial energy costs include the amount of energy consumed to produce, deliver and install the material. The long term energy cost is the economic, ecological, and social costs of continuing to produce and deliver energy to the building for its' use, maintenance, and eventual removal.

The initial embodied energy of a structure is the energy consumed to extract, manufacture, deliver, install, the materials. The life time embodied energy continues to grow with the use, maintenance, and reuse/recycling/disposal of the building materials themselves *and* how the materials and design help minimize the life-time energy consumption of the structure.

Social Costs

Social costs are injury and health of the people producing and transporting the materials and potential health problems of the building occupants if there are problems with the building biology. Globalization has had significant impacts on people both in terms of jobs, skills, and self-sufficiency are lost when manufacturing facilities are closed and the cultural aspects of where new facilities are opened. Aspects of fair trade and labour rights are social costs of global building material manufacturing.

Naturally Occurring Substances

Brush

Brush structures are built entirely from plant parts and were used in primitive cultures such as Native Americans, pygmy peoples in Africa These are built mostly with branches, twigs and leaves, and bark, similar to a beaver's lodge. These were variously named wikiups, lean-tos, and so forth. An extension on the brush building idea is the wattle and daub process in which clay soils or dung, usually cow, are used to fill in and cover a woven brush structure. This gives the structure more thermal mass and strength. Wattle and daub is one of the oldest building techniques. Many older timber frame buildings incorporate wattle and daub as non load bearing walls between the timber frames.

Ice and Snow

Snow and occasionally ice, were used by the Inuit peoples for igloos and snow is used to built a shelter called a quinzhee. Ice has also been used for ice hotels as a tourist attraction in northern climates.

Mud and Clay

Figure: *Sod buildings in Iceland*

Clay based buildings usually come in two distinct types. One being when the walls are made directly with the mud mixture, and the other being walls built by stacking air-dried building blocks or bricks.

Other uses of clay in building is combined with straws to create light clay, wattle and daub, and mud plaster.

Wet-Laid Clay Walls

Wet-laid, or damp, walls are made by using the mud or clay mixture directly without forming blocks and drying them first. The amount of and type of each material in the mixture used leads to different styles of buildings. The deciding factor is usually connected with the quality of the soil being used. Larger amounts of clay are usually employed in building with *cob*, while low-clay soil is usually associated with *sod house* or *sod roof* construction. The other main ingredients include more or less sand/gravel and straw/grasses. *Rammed earth* is both an old and newer take on creating walls, once made by compacting clay soils between planks by hand; nowadays forms and mechanical pneumatic compressors are used.

Soil, and especially clay, provides good thermal mass; it is very good at keeping temperatures at a constant level. Homes built with earth tend to be naturally cool in the summer heat and warm in cold weather. Clay holds heat or cold, releasing it over a period of time like stone. Earthen walls change temperature slowly, so artificially raising or lowering the temperature can use more resources than in say a wood built house, but the heat/coolness stays longer.

Peoples building with mostly dirt and clay, such as cob, sod, and adobe, created homes that have been built for centuries in western and northern Europe, Asia, as well as the rest of the world, and continue to be built, though on a smaller scale. Some of these buildings have remained habitable for hundreds of years.

Structural Clay Blocks and Bricks

Mud-bricks, also known by there Spanish name *adobe* are ancient building materials with evidence dating back thousands of years BC. *Compressed earth blocks* are a more modern type of brick used for building more frequently in industrialized society since the building blocks can be manufactured off site in a centralized location at a brickworks and transported to multiple building locations. These blocks can also be monetized more easily and sold.

Structural mud bricks are almost always made using clay, often clay soil and a binder are the only ingredients used, but other ingredients can include sand, lime, concrete, stone and other binders. The formed or compressed block is then air dried and can be laid dry or with a mortar or clay slip.

Sand

Sand is used with cement, and sometimes lime, to make mortar for masonry work and plaster. Sand is also used as a part of the concrete mix. An important low-cost building material in countries with high sand content soils is the Sandcrete block, which is weaker but cheaper than fired clay bricks.

Stone or Rock

Rock structures have existed for as long as history can recall. It is the longest lasting building material available, and is usually readily available. There are many types of rock throughout the world, all with differing attributes that make them better or worse for particular uses. Rock is a very dense material so it gives a lot of protection too; its main drawback as a material is its weight and awkwardness. Its energy density is also considered a big drawback, as stone is hard to keep warm without using large amounts of heating resources.

Dry-stone walls have been built for as long as humans have put one stone on top of another. Eventually, different forms of mortar were used to hold the stones together, cement being the most commonplace now. The granite-strewn uplands of Dartmoor National Park, United Kingdom, for example, provided ample resources for early settlers. Circular huts were constructed from loose granite rocks throughout the Neolithic and early Bronze Age, and the remains of an estimated 5,000 can still be seen today. Granite continued to be used throughout the Medieval period and into modern times. Slate is another stone type, commonly used as roofing material in the United Kingdom and other parts of the world where it is found.

Stone buildings can be seen in most major cities; some civilizations built entirely with stone such as the Egyptian and Aztec pyramids and the remains of the Inca civilization.

Thatch

Thatch is one of the oldest of building materials known; grass is a good insulator and easily harvested. Many African tribes have lived in homes made completely of grasses and sand year-round. In Europe, thatch roofs on homes were once prevalent but the material fell out of favour as industrialization and improved transport increased the availability of other materials. Today, though, the practice is undergoing a revival. In the Netherlands, for instance, many new buildings have thatched roofs with special ridge tiles on top.

Figure: *Toda tribe hut*

Wood and Timber

Figure: *A wood-framed house under construction in Texas, United States*

Wood has been used as a building material for thousands of years in its natural state. Today, engineered wood is becoming very common in industrialized countries. Wood is a product of trees, and sometimes other fibrous plants, used for construction purposes when cut or pressed into lumber and timber, such as boards, planks and similar materials.

It is a generic building material and is used in building just about any type of structure in most climates. Wood can be very flexible under loads, keeping strength while bending, and is incredibly strong when compressed vertically. There are many differing qualities to the different types of wood, even among same tree species. This means specific species are better suited for various uses than others. And growing conditions are important for deciding quality.

Figure: *The Gliwice Radio Tower (the second tallest wooden structure in the world) in Poland (2012).*

"Timber" is the term used for construction purposes except the term "lumber" is used in the United States. Raw wood (a log, trunk, bole) becomes timber when the wood has been "converted" (sawn, hewn, split) in the forms of minimally-processed logs stacked on top of each other, timber frame construction, and light-frame construction. The main problems with timber structures are fire risk and moisture-related problems.

In modern times softwood is used as a lower-value bulk material, whereas hardwood is usually used for finishings and furniture. Historically timber frame structures were built with oak in western Europe, recently douglas fir has become the most popular wood for most types of structural building.

Many families or communities, in rural areas, have a personal woodlot from which the family or community will grow and harvest trees to build with or sell. These lots are tended to like a garden. This was much more prevalent in pre-industrial times, when laws existed as to the amount of wood one could cut at any one time to ensure there would be a supply of timber for the future, but is still a viable form of agriculture.

Man-Made Substances

Fired Bricks and Clay Blocks

Figure: *A pile of fired bricks.*

Figure: *Clay blocks (sometimes called clay block brick) being laid with an adhesive rather than mortar*

Bricks are made in a similar way to mud-bricks except without the fibrous binder such as straw and are *fired* ("burned" in a brick clamp or kiln) after they have air-dried to permanently harden them. Kiln fired clay bricks are a ceramic material. Fired bricks can be solid or have hollow cavities to aid in drying and make them lighter and easier to transport.

The individual bricks are placed upon each other in courses using mortar. Successive courses being used to build up walls, arches, and other architectural elements. Fired brick walls are usually substantially thinner than cob/adobe while keeping the same vertical strength. They require more energy to create but are easier to transport and store, and are lighter than stone blocks. Romans extensively used fired brick of a shape and type now called Roman bricks. Building with brick gained much popularity in the mid-18th century and 19th centuries. This was due to lower costs with increases in brick manufacturing and fire-safety in the ever crowding cities.

The cinder block supplemented or replaced fired bricks in the late 20th century often being used for the inner parts of masonry walls and by themselves. Structural clay tiles (clay blocks) are clay or terracotta and typically are perforated with holes.

Cement Composites

Cement bonded composites are made of hydrated cement paste that binds wood, particles, or fibres to make pre-cast building components. Various fibreous materials, including paper, fibreglass, and carbon-fibre have been used as binders.

Wood and natural fibres are composed of various soluble organic compounds like carbohydrates, glycosides and phenolics. These compounds are known to retard cement setting. Therefore, before using a wood in making cement bonded composites, its compatibility with cement is assessed.

Wood-cement compatibility is the ratio of a parameter related to the property of a wood-cement composite to that of a neat cement paste. The compatibility is often expressed as a percentage value. To determine wood-cement compatibility, methods based on different properties are used, such as, hydration characteristics, strength, interfacial bond and morphology. Various methods are used by researchers such as the measurement of hydration characteristics of a cement-aggregate mix; the comparison of the mechanical properties of cement-aggregate mixes and the visual assessment of microstructural properties of the wood-cement mixes. It has been found that the hydration test by measuring the change in hydration temperature with time is the most convenient method. Recently, Karade et al. have reviewed these methods of compatibility assessment and suggested a method based on the 'maturity concept' i.e. taking in consideration both time and temperature of cement hydration reaction.

Bricks were laid in lime mortar from the time of the Romans until supplanted by Portland cement mortar in the early 20th century. Cement blocks also sometimes are filled with grout or covered with a parge coat.

Concrete

Concrete is a composite building material made from the combination of aggregate and a binder such as cement. The most common form of concrete is Portland cement concrete, which consists of mineral aggregate (generally gravel and sand), portland cement and water.

Figure: *Falkirk Wheel*

After mixing, the cement hydrates and eventually hardens into a stone-like material. When used in the generic sense, this is the material referred to by the term "concrete".

For a concrete construction of any size, as concrete has a rather low tensile strength, it is generally strengthened using steel rods or bars (known as rebars). This strengthened concrete is then referred to as reinforced concrete. In order to minimise any air bubbles, that would weaken the structure, a vibrator is used to eliminate any air that has been entrained when the liquid concrete mix is poured around the ironwork. Concrete has been the predominant building material in the modern age due to its longevity, formability, and ease of transport. Recent advancements, such as insulating concrete forms, combine the concrete forming and other construction steps (installation of insulation). All materials must be taken in required proportions as described in standards.

Fabric

The tent is the home of choice among nomadic groups all over the world. Two well-known types include the conical teepee and the circular yurt. The tent has been revived as a major construction technique with the development of tensile architecture and synthetic fabrics. Modern buildings can be made of flexible material such as fabric membranes, and supported by a system of steel cables, rigid or internal, or by air pressure.

Foam

Figure: *Foamed plastic sheet to be used as backing for firestop mortar at CIBC bank in Toronto.*

Recently, synthetic polystyrene or polyurethane foam has been used in combination with structural materials, such as concrete. It is lightweight, easily shaped, and an excellent insulator. Foam is usually used as part of a structural insulated panel, wherein the foam is sandwiched between wood or cement or insulating concrete forms.

Glass

Glassmaking is considered an art form as well as an industrial process or material.

Clear windows have been used since the invention of glass to cover small openings in a building. Glass panes provided humans with the ability to both let light into rooms while at the same time keeping inclement weather outside.

Glass is generally made from mixtures of sand and silicates, in a very hot fire stove called a kiln, and is very brittle. Additives are often included the mixture used to produce glass with shades of colours or various characteristics (such as bulletproof glass or light emittance).

The use of glass in architectural buildings has become very popular in the modern culture. Glass "curtain walls" can be used to cover the entire facade of a building, or it can be used to span over a wide roof structure in a "space frame". These uses though require some sort of frame to hold sections of glass together, as glass by itself is too brittle and would require an overly large kiln to be used to span such large areas by itself.

Glass bricks were invented in the early 20th century.

Gypcrete

Gypcrete is a mixture of gypsum plaster and fibreglass rovings. Although plaster and fibres fiborous plaster have been used for many years, especially for ceilings, it was not until the early 1990s that serious studies of the strength and qualities of a walling system Rapidwall, using a mixture of gypsum plaster and 300mm plus fibreglass rovings, were investigated. It was discovered, through testing at the University of Adelaide, that these walls had significant, load bearing, shear and lateral resistance together with earthquake-resistance, fire-resistance, and thermal properties. With an abundance of gypsum (naturally occurring and by-product chemical FGD and phospho gypsums) available worldwide, gypcrete-based building products, which are fully recyclable, offer significant environmental benefits.

Metal

Metal is used as structural framework for larger buildings such as skyscrapers, or as an external surface covering. There are many types of metals used for building. Metal figures quite prominently in prefabricated structures such as the Quonset hut, and can be seen used in most cosmopolitan cities. It requires a great deal of human labour to produce metal, especially in the large amounts needed for the building industries. Corrosion is metal's prime enemy when it comes to longevity.

- Steel is a metal alloy whose major component is iron, and is the usual choice for metal structural building materials. It is strong, flexible, and if refined well and/or treated lasts a long time.
- The lower density and better corrosion resistance of aluminium alloys and tin sometimes overcome their greater cost.
- Copper is a valued building material because of its advantageous properties. These include corrosion resistance, durability, low thermal movement, light weight, radio frequency shielding, lightning protection, sustainability, recyclability, and a wide range of finishes. Copper is incorporated into roofing, flashing, gutters, downspouts, domes, spires, vaults, wall cladding, building expansion joints, and indoor design elements.

- Other metals used include chrome, gold, silver, and titanium. Titanium can be used for structural purposes, but it is much more expensive than steel. Chrome, gold, and silver are used as decoration, because these materials are expensive and lack structural qualities such as tensile strength or hardness.

Plastics

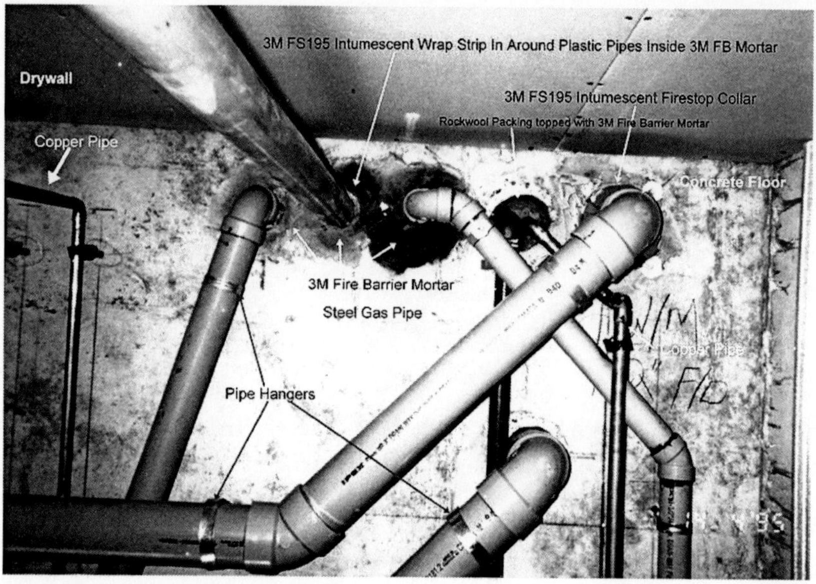

Figure: *Plastic pipes penetrating a concrete floor in a Canadian highrise apartment building*

The term “plastics” covers a range of synthetic or semi-synthetic organic condensation or polymerization products that can be molded or extruded into objects, films, or fibres. Their name is derived from the fact that in their semi-liquid state they are malleable, or have the property of plasticity. Plastics vary immensely in heat tolerance, hardness, and resiliency. Combined with this adaptability, the general uniformity of composition and lightness of plastics ensures their use in almost all industrial applications today.

Papers and Membranes

Building papers and membranes are used for many reasons in construction. One of the oldest building papers is red rosin paper which was known to be in use before 1850 and was used as an underlayment in exterior walls, roofs, and floors and for protecting a jobsite during construction. Tar paper was invented late in the 19th century and was used for similar purposes as rosin paper and for gravel roofs. Tar paper has largely fallen out of use supplanted by asphalt felt paper. Felt paper has been supplanted in some uses by synthetic underlayments, particularly in roofing by synthetic underlayments and siding by housewraps.

There are a wide variety of damp proofing and waterproofing membranes used for roofing, basement waterproofing, and geomembranes.

Ceramics

Fired clay bricks have been used since the time of the Romans. Special tiles are used for roofing, siding, flooring, ceilings, pipes, flue liners, and more.

Building Products

In the market place the term "building products" often refers to ready-made particles/sections, made from various materials, that are fitted in architectural hardware and decorative hardware parts of a building. The list of building products excludes the building materials used to construct the building architecture and supporting fixtures, like windows, doors, cabinets, etc. Building products, rather, support and make building materials work in a modular fashion.

"Building products" may also refer to items used to put such hardware together, such as caulking, glues, paint, and anything else bought for the purpose of constructing a building.

Testing and Certification

- ASTM International
- UL (safety organisation)
- ETL SEMKO — Building Product Testing Laboratory in the USA, part of Intertek, based in London
- NTA Inc— Building Product Certification Agency based in Nappanee, Indiana, USA

Nano Metallurgy

Metallurgy is a domain of materials science and materials engineering that studies the physical and chemical behavior of metallic elements, their intermetallic compounds, and their mixtures, which are called alloys. Metallurgy is also the technology of metals: the way in which science is applied to the production of metals, and the engineering of metal components for use in products for consumers and manufacturers. The production of metals involves the processing of ores to extract the metal they contain, and the mixture of metals, sometimes with other elements, to produce alloys. Metallurgy is distinguished from the craft of metalworking.

Metallurgy is subdivided into *ferrous metallurgy* (sometimes also known as black metallurgy) and *non-ferrous metallurgy* or *coloured metallurgy*. Ferrous metallurgy involves processes and alloys based on iron while non-ferrous metallurgy involves processes and alloys based on other metals. The production of ferrous metals accounts for 95 percent of world metal production.

Microstructure

Metallurgists study the microscopic and macroscopic properties using metallography, a technique invented by Henry Clifton Sorby. In metallography, an alloy of interest is ground flat and polished to a mirror finish. The sample can then be etched to reveal the microstructure and macrostructure of the metal. The sample is then examined in an optical or electron microscope, and the image contrast provides details on the composition, mechanical properties, and processing history.

Crystallography, often using diffraction of x-rays or electrons, is another valuable tool available to the modern metallurgist. Crystallography allows identification of unknown materials and reveals the crystal structure of the sample. Quantitative crystallography can be used to calculate the amount of phases present as well as the degree of strain to which a sample has been subjected.

Nano-Metallurgy for High Temperature Metals

The inberent properties of metals are revealed sometimes by the ultra-purifucation. For example, the properties of iron with 99.999% purity after the ultra - purified in the ultra-high vacuum of 10-10Torr is rust free in the air at room temperature, deformable in the liquid nitrogen, recrystallized at 300°C, transformed with very high mobility of the α ÿãinterface. In the ultra-purified iron-carbon alloy, granular cementite is more stable than planer cementite. After the ultra-purification of iron-chromium alloy, the inberent properties are revealed.

At the beginning of the 21st century, the rapid progress in the foremost industries such as the energy industry is demanded. The development of innovative metallic materials is important for the severe demand. In the laboratory, the funda-mental research on the ultra-purification of metals such as iron, chromium, nickel, titanium, cobalt and their alloys and on the clarication of the inberent properties of those metals as well as to the practical use is going on.

Neutron Diffraction

Neutron diffraction or elastic neutron scattering is the application of neutron scattering to the determination of the atomic and/or magnetic structure of a material. A sample to be examined is placed in a beam of thermal or cold neutrons to obtain a diffraction pattern that provides information of the structure of the material. The technique is similar to X-ray diffraction but due to their different scattering properties, neutrons and X-rays provide complementary information.

Instrumental and Sample Requirements

The technique requires a source of neutrons. Neutrons are usually produced in a nuclear reactor or spallation source. At a research reactor, other components are needed, including a crystal monochromators as well as filters to select the desired neutron wavelength. Some parts of the setup may also be movable. At a spallation source, the time of flight technique is used to sort the energies of the incident neutrons (higher energy neutrons are faster), so no monochromator is needed, but rather a series of aperture elements synchronized to filter neutron pulses with the desired wavelength.

The technique is most commonly performed as powder diffraction, which only requires a polycrystalline powder. For single crystal work, the crystals must be much larger than those used in X-ray crystallography. It is common to use crystals that are about 1 mm^3.

Summarizing, the main disadvantage to neutron diffraction is the requirement for a nuclear reactor. For single crystal work, the technique requires relatively large crystals, which are usually challenging to grow. The main advantages to the technique are many - sensitivity to light atoms, ability to distinguish isotopes, absence of radiation damage.

Nuclear Scattering

Like all quantum particles, neutrons can exhibit wave phenomena typically associated with light or sound. Diffraction is one of these phenomena; it occurs when waves encounter obstacles whose size is comparable with the wavelength. If the wavelength of a quantum particle is short enough, atoms or their nuclei can serve as diffraction obstacles. When a beam of neutrons emanating from a reactor is slowed down and selected properly by their speed, their wavelength lies near one angstrom (0.1 nanometre), the typical separation between atoms in a solid material. Such a beam can then be used to perform a diffraction experiment. Impinging on a crystalline sample it will scatter under a limited number of well-defined angles according to the same Bragg's law that describes X-ray diffraction.

Neutrons and X-rays interact with matter differently. X-rays interact primarily with the electron cloud surrounding each atom. The contribution to the diffracted x-ray intensity is therefore larger for atoms with larger atomic number (Z). On the other hand, neutrons interact directly with the *nucleus* of the atom, and the contribution to the diffracted intensity depends on each isotope; for example, regular hydrogen and deuterium contribute differently. It is also often the case that light (low Z) atoms contribute strongly to the diffracted intensity even in the presence of large Z atoms. The scattering length varies from isotope to isotope rather than linearly with the atomic number. An element like vanadium is a strong scatterer of X-rays, but its nuclei hardly scatter neutrons, which is why it often used as a container material. Non-magnetic neutron diffraction is directly sensitive to the positions of the nuclei of the atoms.

Unlike X-rays, neutrons scatter mostly from to the nuclei of the atoms, which are tiny. Furthermore, there is no need for an atomic form factor to describe the shape of the electron cloud of the atom and the scattering power of an atom does not fall off with the scattering angle as it does for X-rays. Diffractograms therefore can show strong well defined diffraction peaks even at high angles, particularly if the experiment is done at low temperatures. Many neutron sources are equipped with liquid helium cooling systems that allow data collection at temperatures down to 4.2 K. The superb high angle (i.e. high *resolution*) information means that the atomic positions in the structure can be determined with high precision. On the other hand, Fourier maps (and to a lesser extent difference Fourier maps) derived from neutron data suffer from series termination errors, sometimes so much that the results are meaningless.

Magnetic Scattering

Although neutrons are uncharged, they carry a spin, and therefore interact with magnetic moments, including those arising from the electron cloud around an atom. Neutron diffraction can therefore reveal the microscopic magnetic structure of a material.

Magnetic scattering does require an atomic form factor as it is caused by the much larger electron cloud around the tiny nucleus. The intensity of the magnetic contribution to the diffraction peaks will therefore dwindle towards higher angles.

Uses

Neutron diffraction can be used to determine the static structure factor of gases, liquids or amorphous solids. Most experiments, however, aim at the structure of crystalline solids, making neutron diffraction an important tool of crystallography. Neutron diffraction is closely related to X-ray powder diffraction. In fact the single crystal version of the technique is less commonly used because currently available neutron sources require relatively large samples and large single crystals are hard or impossible to come by for most materials. Future developments, however, may well change this picture. Because the data is typically a 1D powder diffractogram they are usually processed using Rietveld refinement. In fact the latter found its origin in neutron diffraction (at Petten in the Netherlands) and was later extended for use in X-ray diffraction.

One practical application of elastic neutron scattering/diffraction is that the lattice constant of metals and other crystalline materials can be very accurately measured. Together with an accurately aligned micropositioner a map of the lattice constant through the metal can be derived. This can easily be converted to the stress field experienced by the material. This has been used to analyse stresses in aerospace and automotive components to give just two examples. This technique has led to the development of dedicated stress diffractometres, such as the ENGIN-X instrument at the ISIS neutron source.

Neutron Diffraction can also be employed to give insight into the 3D structure any material that diffracts.

Hydrogen, Null-Scattering and Contrast Variation

Neutron diffraction can be used to establish the structure of low atomic number materials like proteins and surfactants much more easily with lower flux than at a synchrotron radiation source. This is because some low atomic number materials have a higher cross section for neutron interaction than higher atomic weight materials.

One major advantage of neutron diffraction over X-ray diffraction is that the latter is rather insensitive to the presence of hydrogen (H) in a structure, whereas the nuclei ^1H and ^2H (i.e. Deuterium, D) are strong scatterers for neutrons. The greater scattering power of protons and deuterons means that the position of hydrogen in a crystal and its thermal motions can be determined with greater precision by neutron diffraction. The structures of metal hydride complexes, e.g., Mg_2FeH_6 have been assessed by neutron diffraction.

The neutron scattering lengths b_H = -3.7406(11) fm and b_D = 6.671(4) fm, for H and D respectively, have opposite sign, which allows the technique to distinguish them. In fact there is a particular isotope ratio for which the contribution of the element would cancel, this is called null-scattering.

It is undesirable to work with the relatively high concentration of H in a sample. The scattering intensity by H-nuclei has a large inelastic component, which creates a large continuous background that is more or less independent of scattering angle. The elastic pattern typically consists of sharp Bragg reflections if the sample is crystalline. They tend to drown in the inelastic background. This is even more serious when the technique is used for the study of liquid structure. Nevertheless, by preparing samples with different isotope ratios it is possible to vary the scattering contrast enough to highlight one element in an otherwise complicated structure. The variation of other elements is possible but usually rather expensive. Hydrogen is inexpensive and particularly interesting because it plays an exceptionally large role in biochemical structures and is difficult to study structurally in other ways.

Pair Distribution Function

The pair distribution function (PDF) describes the distribution of distances between pairs of particles contained within a given volume. Mathematically, if a and b are two particles in a fluid, the PDF of b with respect to a, denoted by $g_{ab}(\vec{r})$ is the probability of finding the particle b at the distance $\vec{r}$ from a, with a taken as the origin of coordinates.

The pair distribution function is used to describe the distribution of objects within a medium (for example, oranges in a crate or nitrogen molecules in a gas cylinder). If the medium is homogeneous (i.e. every spatial location has identical properties), then there is an equal probability density for finding an object at any position $\vec{r}$:

$$p(\vec{r}) = 1/V\ ,$$

where V is the volume of the container. On the other hand, the likelihood of finding *pairs of objects* at given positions (i.e. the two-body probability density) is not uniform. For example, pairs of hard balls must be separated by at least the diameter of a ball. The pair distribution function $g(\vec{r},\vec{r}')$ is obtained by scaling the two-body probability density function by the total number of objects N and the size of the container:

$$g(\vec{r},\vec{r}') = p(\vec{r},\vec{r}')V^2 \frac{N-1}{N}\ .$$

In the common case where the number of objects in the container is large, this simplifies to give:

$$g(\vec{r},\vec{r}') \approx p(\vec{r},\vec{r}')V^2\ .$$

Simple Models and General Properties

The simplest possible pair distribution function assumes that all object locations are mutually independent, giving:

$$g(\vec{r}) = 1\ ,$$

where $\vec{r}$ is the separation between a pair of objects. However, this is inaccurate in the case of hard objects as discussed above, because it does not account for the minimum separation required between objects. The hole-correction (HC) approximation provides a better model:

$$g(r) = \begin{cases} 0, & r < b, \\ 1, & r \geq b \end{cases},$$

where b is the diameter of one of the objects.

Although the HC approximation gives a reasonable description of sparsely packed objects, it breaks down for dense packing. This may be illustrated by considering a box completely filled by identical hard balls so that each ball touches its neighbours.

In this case, every pair of balls in the box is separated by a distance of exactly $r = nb$ where n is a positive whole number. The pair distribution for a volume completely filled by hard spheres is therefore a set of Dirac delta functions of the form:

$$g(r) = \sum_i \delta(r - ib).$$

Finally, it may be noted that a pair of objects which are separated by a large distance have no influence on each other's position (provided that the container is not completely filled). Therefore,

$$\lim_{r \to \infty} g(r) = 1.$$

In general, a pair distribution function will take a form somewhere between the sparsely packed (HC approximation) and the densely packed (delta function) models, depending on the packing density f.

Radial Pair Distributions

Of special practical importance is the radial pair distribution function, which is independent of orientation. It is a major descriptor for the atomic structure of amorphous materials (glasses, polymers) and liquids.

The radial PDF can be calculated directly from physical measurements like light scattering or x-ray powder diffraction through the use of Fourier Transform.

In Statistical Mechanics the PDF is given by the expression:

$$g_{ab}(r) = \frac{1}{N_a N_b} \sum_{i=1}^{N_a} \sum_{j=1}^{N_b} \langle \delta(|\mathbf{r}_{ij}| - r) \rangle$$

The Diffpy project is used to match crystal structures with PDF data derived from X-ray or neutron diffraction data.

Solid State Chemistry

Solid-state chemistry, also sometimes referred to as materials chemistry, is the study of the synthesis, structure, and properties of solid phase materials, particularly, but not necessarily exclusively of, non-molecular solids. It therefore has a strong overlap with solid-state physics, mineralogy, crystallography, ceramics, metallurgy, thermodynamics, materials science and electronics with a focus on the synthesis of novel materials and their characterization.

Because of its direct relevance to products of commerce, solid state inorganic chemistry has been strongly driven by technology. Progress in the field has often been fuelled by the demands of industry, well ahead of purely academic curiosity. Applications discovered in the 20th century include zeolite and platinum-based catalysts for petroleum processing in the 1950s, high-purity silicon as a core component of microelectronic devices in the 1960s, and "high temperature" superconductivity in the 1980s. The invention of X-ray crystallography in the early 1900s by William Lawrence Bragg enabled further innovation. Our understanding of how reactions proceed at the atomic level in the solid state was advanced considerably by Carl Wagner's work on oxidation rate theory, counter diffusion of ions, and defect chemistry. Because of this, he has sometimes been referred to as the *father of solid state chemistry*.

Synthetic Methods

Given the diversity of solid state compounds, an equally diverse array of methods are used for their preparation. For organic materials, such as charge transfer salts, the methods operate near room temperature and are often similar to the techniques of organic synthesis. Redox reactions are sometimes conducted by electrocrystallisation, as illustrated by the preparation of the Bechgaard salts from tetrathiafulvalene.

Oven Techniques

For thermally robust materials, high temperature methods are often employed. For example, bulk solids are prepared using tube furnaces, which allow reactions to be conducted up to ca. 1100 °C. Special equipment e.g. ovens consisting of a tantalum tube through which an electric current is passed can be used for even higher temperatures up to 2000 °C. Such high temperatures are at times required to induce diffusion of the reactants, but this depends strongly on the system studied. Some solid state reactions already proceed at temperatures as low as 100 °C.

Melt Methods

One method often employed is to melt the reactants together and then later anneal the solidified melt. If volatile reactants are involved the reactants are often put in an ampoule that is evacuated -often while keeping the reactant mixture cold e.g. by keeping the bottom of the ampoule in liquid nitrogen- and then sealed. The sealed ampoule is then put in an oven and given a certain heat treatment.

Solution Methods

It is possible to use solvents to prepare solids by precipitation or by evaporation. At times the solvent is used hydrothermally, i.e. under pressure at temperatures higher than the normal boiling point.

A variation on this theme is the use of flux methods, where a salt of relatively low melting point is added to the mixture to act as a high temperature solvent in which the desired reaction can take place.

Gas Reactions

Many solids react readily with reactive gas species like chlorine, iodine, oxygen etc. Others form adducts with other gases, e.g. CO or ethylene. Such reactions are often carried out in a tube that is open ended on both sides and through which the gas is passed. A variation of this is to let the reaction take place inside a measuring device such as a TGA. In that case stoichiometric information can be obtained during the reaction, which helps identify the products.

A special case of a gas reaction is a chemical transport reaction. These are often carried out in a sealed ampoule to which a small amount of a transport agent, e.g. iodine is added. The ampoule is then placed in a zone oven.

This is essentially two tube ovens attached to each other which allows a temperature gradient to be imposed. Such a method can be used to obtain the product in the form of single crystals suitable for structure determination by X-ray diffraction.

Chemical vapor deposition is a high temperature method that is widely employed for the preparation of coatings and semiconductors from molecular precursors.

Air and Moisture Sensitive Materials

Many solids are hygroscopic and/or oxygen sensitive. Many halides e.g. are very 'thirsty' and can only be studied in their anhydrous form if they are handled in a glove box filled with dry (and/or oxygen-free) gas, usually nitrogen.

Characterization

New Phases, Phase Diagrams, Structures

The synthetic methodology and the characterization of the product often go hand in hand in the sense that not one but a series of reaction mixtures are prepared and subjected to heat treatment. The stoichiometry is typically *varied* in a systematic way to find which stoichiometries will lead to new solid compounds or to solid solutions between known ones. A prime method to characterize the reaction products is powder diffraction, because many solid state reactions will produce polycristalline ingots or powders. Powder diffraction will facilitate the identification of known phases in the mixture. If a pattern is found that is not known in the diffraction data libraries an attempt can be made to index the pattern, i.e. to identify the symmetry and the size of the unit cell. (If the product is not crystalline the characterization is typically much more difficult.)

Once the unit cell of a new phase is known, the next step is to establish the stoichiometry of the phase. This can be done in a number of ways. Sometimes the composition of the original mixture will give a clue, if one finds only one product -a single powder pattern- or if one was trying to make a phase of a certain composition by analogy to known materials but this is rare. Often considerable effort in refining the synthetic methodology is required to obtain a pure sample of the new material. If it is possible to separate the product from the rest of the reaction mixture elemental analysis can be used. Another ways involves SEM and the generation of characteristic X-rays in the electron beam. The easiest way to solve the structure is by using single crystal X-ray diffraction.

The latter often requires *revisiting* and refining the preparative procedures and that is linked to the question which phases are stable at what composition and what stoichiometry. In other words what does the phase diagram looks like. An important tool in establishing this is thermal analysis techniques like DSC or DTA and increasingly also, thanks to the advent of synchrotrons temperature-dependent powder diffraction. Increased knowledge of the phase relations often leads to further refinement in synthetic procedures in an iterative way. New phases are thus characterized by their melting points and their stoichiometric domains. The latter is important for the many solids that are non-stoichiometric compounds. The cell parameters obtained from XRD are particularly helpful to characterize the homogeneity ranges of the latter.

Further Characterization

In many -but certainly not all- cases new solid compounds are further characterized by a variety of techniques that straddle the fine line that (hardly) separates solid-state chemistry from solid-state physics.

Optical Properties

For non-metallic materials it is often possible to obtain UV/VIS spectra. In the case of semiconductors that will give an idea of the band gap.

Texture (Crystalline)

In materials science, texture is the distribution of crystallographic orientations of a polycrystalline sample. A sample in which these orientations are fully random is said to have no texture. If the crystallographic orientations are not random, but have some preferred orientation, then the sample has a weak, moderate or strong texture. The degree is dependent on the percentage of crystals having the preferred orientation. Texture is seen in almost all engineered materials, and can have a great influence on material properties. Also, geologic rocks show texture due to their thermo-mechanic history of formation processes.

One extreme case is a complete lack of texture: a solid with perfectly random crystallite orientation will have isotropic properties at length scales sufficiently larger than the size of the crystallites. The opposite extreme is a perfect single crystal, which has anisotropic properties by geometric necessity.

Characterization and Representation

Texture can be determined by various methods. Some of them allow a quantitative analysis of the texture; others are only qualitative. Among the quantitative techniques, the most widely used is X-ray diffraction using texture goniometres, followed by EBSD-method (electron backscatter diffraction) in Scanning Electron Microscopes. Qualitative analysis can be done by Laue photography, simple X-ray diffraction or with the polarized microscope. Neutron and synchrotron high-energy X-ray diffraction allow access to textures of bulk material and in-situ analysis, whereas laboratory x-ray diffraction instruments are more appropriate for thin film textures.

Texture is often represented using a pole figure, in which a specified crystallographic axis (or pole) from each of a representative number of crystallites is plotted in a stereographic projection, along with directions relevant to the material's processing history. These directions define the so-called sample reference frame and are, because the investigation of textures started from the cold working of metals, usually referred to as the rolling direction *RD*, the transverse direction *TD* and the normal direction *ND*. For drawn metal wires the cylindrical fibre axis turned out as the sample direction around which preferred orientation is typically observed.

Common Textures

There are several textures that are commonly found in processed materials. They are named either by the scientist that discovered them, or by the material they are most found in. These are given in miller indices for simplification purposes.

- Cube component: (001)[100]
- Brass component: (110)[-112]
- Copper component: (112)[11-1]
- S component: (123)[63-4]

Origins

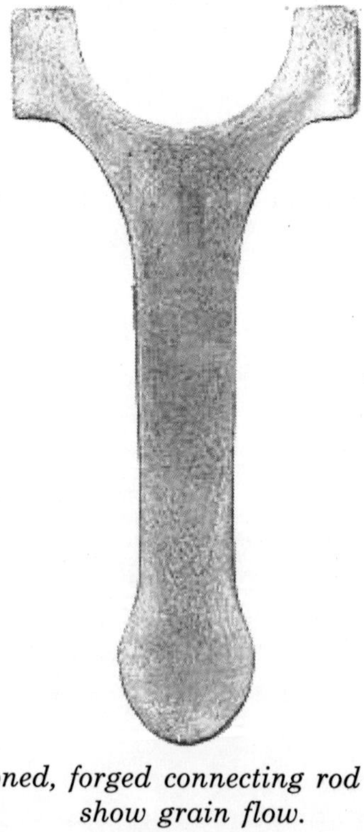

Figure: *Scan of sectioned, forged connecting rod that has been etched to show grain flow.*

In wire and fibre, all crystals tend to have nearly identical orientation in the axial direction, but nearly random radial orientation. The most familiar exceptions to this rule are fibreglass, which has no crystal structure, and carbon fibre, in which the crystalline anisotropy is so great that a good-quality filament will be a distorted single crystal with approximately cylindrical symmetry (often compared to a jelly roll). Single-crystal fibres are also not uncommon.

The making of metal sheet often involves compression in one direction and, in efficient rolling operations, tension in another, which can orient crystallites in both axes by a process known as grain flow. However, cold work destroys much of the crystalline order, and the new crystallites that arise with annealing usually have a different texture. Control of texture is extremely important in the making of silicon steel sheet for transformer cores (to reduce magnetic hysteresis) and of aluminium cans (since deep drawing requires extreme and relatively uniform plasticity).

Texture in ceramics usually arises because the crystallites in a slurry have shapes that depend on crystalline orientation, often needle- or plate-shaped. These particles align themselves as water leaves the slurry, or as clay is formed.

Casting or other fluid-to-solid transitions (i.e., thin-film deposition) produce textured solids when there is enough time and activation energy for atoms to find places in existing crystals, rather than condensing as an amorphous solid or starting new crystals of random orientation. Some facets of a crystal (often the close-packed planes) grow more rapidly than others, and the crystallites for which one of these planes faces in the direction of growth will usually out-compete crystals in other orientations. In the extreme, only one crystal will survive after a certain length: this is exploited in the Czochralski process (unless a seed crystal is used) and in the casting of turbine blades and other creep-sensitive parts.

Texture and Materials Properties

Material properties such as strength, chemical reactivity, stress corrosion cracking resistance, weldability, deformation behavior, resistance to radiation damage, and magnetic susceptibility can be highly dependent on the material's texture and related changes in microstructure. In many materials, properties are texture-specific, and development of unfavourable textures when the material is fabricated or in use can create weaknesses that can initiate or exacerbate failures. Parts can fail to perform due to unfavourable textures in their component materials. Failures can correlate with the crystalline textures formed during fabrication or use of that component. Consequently, consideration of textures that are present in and that could form in engineered components while in use can be a critical when making decisions about the selection of some materials and methods employed to manufacture parts with those materials. When parts fail during use or abuse, understanding the textures that occur within those parts can be crucial to meaningful interpretation of failure analysis data.

Thin Film Textures

Pronounced textures occur in thin films. Modern technological devices to a large extent rely on polycrystalline thin films with thickness in the nanometre and micrometre range. This holds, for instance, for all microelectronic and most optoelectronic systems or sensoric and superconducting layers. Most thin film textures may be categorized into two different types: (1) for so-called fibre textures the orientation of a certain lattice plane is preferentially parallel to the substrate plane. (2) In contrast, in biaxial textures also the in-plane orientation of crystallites becomes locked with respect to the sample. The latter phenomenon is accordingly observed in nearly epitaxial growth processes, where the crystallographic axes of the layer tend to align along those of the substrate.

Tailoring the texture on demand became an important task in thin film technology. In the case of oxide compounds intended for transparent conducting films or surface acoustic wave (SAW) devices, for instance, the polar axis should be aligned along the substrate normal. Another example is given by cables from high-temperature superconductors that are being developed as oxide multilayer systems deposited on metallic ribbons. The adjustment of the biaxial texture in $YBa_2Cu_3O_7$ layers turned out as the decisive prerequisite for achieving sufficiently large critical currents.

The degree of texture is often subjected to an evolution during thin film growth and the most pronounced textures are only obtained after the layer has achieved a certain thickness. Thin film growers thus require information about the texture profile or the texture gradient in order to optimize the deposition process. The determination of texture gradients by x-ray scattering, however, is not straightforward, because different depths of a specimen contribute to the signal. Techniques that allow for the adequate deconvolution of diffraction intensity were developed only recently.

5

X-Ray Crystallography

X-ray crystallography is a tool used for determining the atomic and molecular structure of a crystal, in which the crystalline atoms cause a beam of incident X-rays to diffract into many specific directions. By measuring the angles and intensities of these diffracted beams, a crystallographer can produce a three-dimensional picture of the density of electrons within the crystal. From this electron density, the mean positions of the atoms in the crystal can be determined, as well as their chemical bonds, their disorder and various other information.

Since many materials can form crystals—such as salts, metals, minerals, semiconductors, as well as various inorganic, organic and biological molecules—X-ray crystallography has been fundamental in the development of many scientific fields. In its first decades of use, this method determined the size of atoms, the lengths and types of chemical bonds, and the atomic-scale differences among various materials, especially minerals and alloys.

The method also revealed the structure and function of many biological molecules, including vitamins, drugs, proteins and nucleic acids such as DNA. X-ray crystallography is still the chief method for characterizing the atomic structure of new materials and in discerning materials that appear similar by other experiments. X-ray crystal structures can also account for unusual electronic or elastic properties of a material, shed light on chemical interactions and processes, or serve as the basis for designing pharmaceuticals against diseases.

In an X-ray diffraction measurement, a crystal is mounted on a goniometre and gradually rotated while being bombarded with X-rays, producing a diffraction pattern of regularly spaced spots known as *reflections*. The two-dimensional images taken at different rotations are converted into a three-dimensional model of the density of electrons within the crystal using the mathematical method of Fourier transforms, combined with chemical data known for the sample. Poor resolution (fuzziness) or even errors may result if the crystals are too small, or not uniform enough in their internal makeup.

X-ray crystallography is related to several other methods for determining atomic structures. Similar diffraction patterns can be produced by scattering electrons or neutrons, which are likewise interpreted as a Fourier transform. If single crystals of sufficient size cannot be obtained, various other X-ray methods can be applied to obtain less detailed information; such methods include fibre diffraction, powder diffraction and small-angle X-ray scattering (SAXS). If the material under investigation is only available in the form of nanocrystalline powders or suffers from poor crystallinity, the methods of electron crystallography can be applied for determining the atomic structure. For all above mentioned X-ray diffraction methods, the scattering is elastic; the scattered X-rays have the same wavelength as the incoming X-ray. By contrast, *inelastic* X-ray scattering methods are useful in studying excitations of the sample, rather than the distribution of its atoms.

Early Scientific History of Crystals and X-Rays

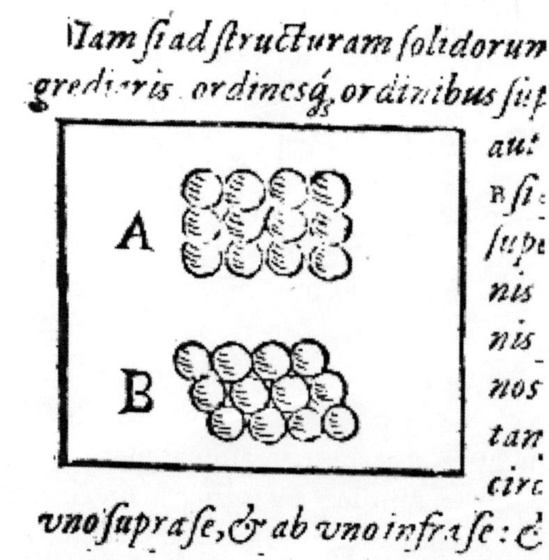

Figure: *Drawing of square (Figure above) and hexagonal (Figure below) packing from Kepler's work,* Strena seu de Nive Sexangula.

Crystals have long been admired for their regularity and symmetry, but they were not investigated scientifically until the 17th century. Johannes Kepler hypothesized in his work *Strena seu de Nive Sexangula* (A New Year's Gift of Hexagonal Snow) (1611) that the hexagonal symmetry of snowflake crystals was due to a regular packing of spherical water particles.

Figure: *As shown by X-ray crystallography, the hexagonal symmetry of snowflakes results from the tetrahedral arrangement of hydrogen bonds about each water molecule. The water molecules are arranged similarly to the silicon atoms in the tridymite polymorph of* SiO_2*. The resulting crystal structure has hexagonal symmetry when viewed along a principal axis.*

Crystal symmetry was first investigated experimentally by Danish scientist Nicolas Steno (1669), who showed that the angles between the faces are the same in every exemplar of a particular type of crystal, and by René Just Haüy (1784), who discovered that every face of a crystal can be described by simple stacking patterns of blocks of the same shape and size.

Hence, William Hallowes Miller in 1839 was able to give each face a unique label of three small integers, the Miller indices which are still used today for identifying crystal faces. Haüy's study led to the correct idea that crystals are a regular three-dimensional array (a Bravais lattice) of atoms and molecules; a single unit cell is repeated indefinitely along three principal directions that are not necessarily perpendicular. In the 19th century, a complete catalog of the possible symmetries of a crystal was worked out by Johan Hessel, Auguste Bravais, Evgraf Fedorov, Arthur Schönflies and (belatedly) William Barlow. From the available data and physical reasoning, Barlow proposed several crystal structures in the 1880s that were validated later by X-ray crystallography; however, the available data were too scarce in the 1880s to accept his models as conclusive.

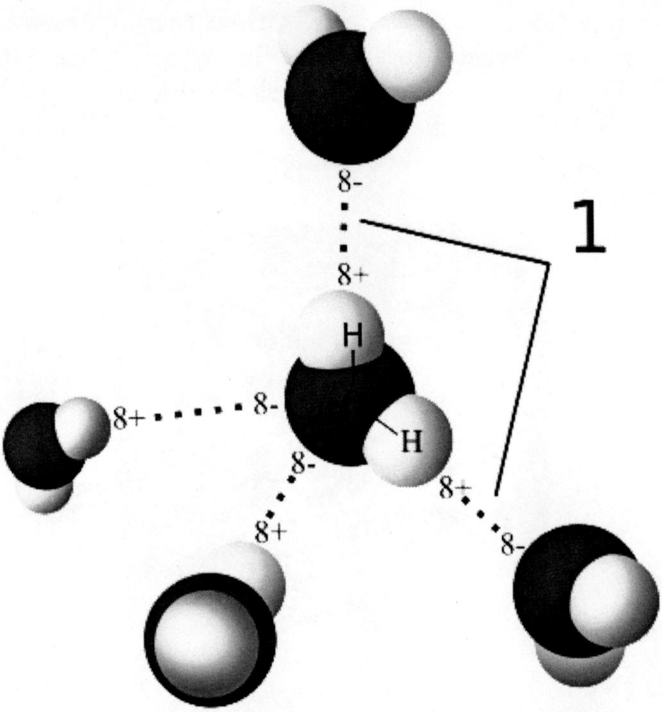

Figure: *X-ray crystallography shows the arrangement of water molecules in ice, revealing the hydrogen bonds (1) that hold the solid together. Few other methods can determine the structure of matter with such precision (*resolution*).*

X-rays were discovered by Wilhelm Conrad Röntgen in 1895, just as the studies of crystal symmetry were being concluded. Physicists were initially uncertain of the nature of X-rays, although it was soon suspected (correctly) that they were waves of electromagnetic radiation, in other words, another form of light. At that time, the wave model of light—specifically, the Maxwell theory of electromagnetic radiation—was well accepted among scientists, and experiments by Charles Glover Barkla showed that X-rays exhibited phenomena associated with electromagnetic waves, including transverse polarization and spectral lines akin to those observed in the visible wavelengths. Single-slit experiments in the laboratory of Arnold Sommerfeld suggested the wavelength of X-rays was about 1 angstrom. However, X-rays are composed of photons, and thus are not only waves of electromagnetic radiation but also exhibit particle-like properties. The photon concept was introduced by Albert Einstein in 1905, but it was not broadly accepted until 1922, when Arthur Compton confirmed it by the scattering of X-rays from electrons. Therefore, these particle-like properties of X-rays, such as their ionization of gases, caused William Henry Bragg to argue in 1907 that X-rays were *not* electromagnetic radiation. Nevertheless, Bragg's view was not broadly accepted and the observation of X-ray diffraction by Max von Laue in 1912 confirmed for most scientists that X-rays were a form of electromagnetic radiation.

X-Ray Analysis of Crystals

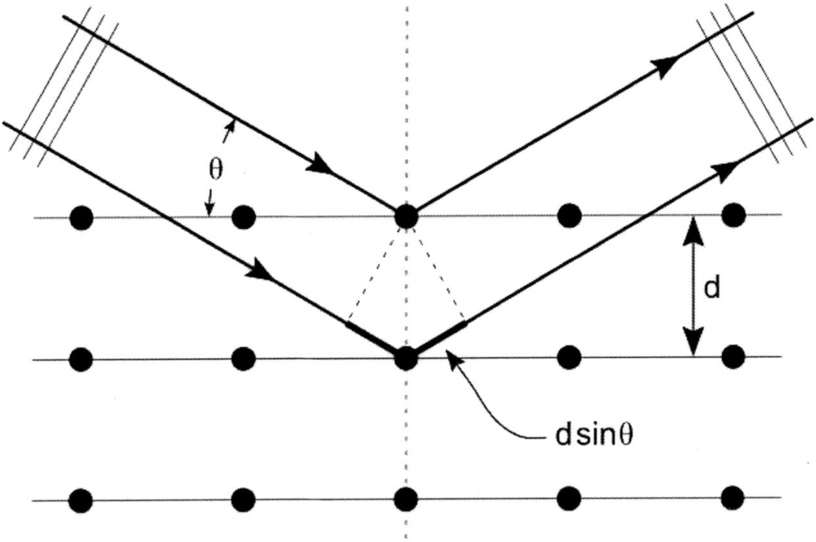

Figure: *The incoming beam (coming from upper left) causes each scatterer to re-radiate a small portion of its intensity as a spherical wave. If scatterers are arranged symmetrically with a separation* d, *these spherical waves will be in sync (add constructively) only in directions where their path-length difference 2*d *sin θ equals an integer multiple of the wavelength λ. In that case, part of the incoming beam is deflected by an angle 2θ, producing a* reflection *spot in the diffraction pattern.*

Crystals are regular arrays of atoms, and X-rays can be considered waves of electromagnetic radiation. Atoms scatter X-ray waves, primarily through the atoms' electrons. Just as an ocean wave striking a lighthouse produces secondary circular waves emanating from the lighthouse, so an X-ray striking an electron produces secondary spherical waves emanating from the electron. This phenomenon is known as elastic scattering, and the electron (or lighthouse) is known as the *scatterer*. A regular array of scatterers produces a regular array of spherical waves. Although these waves cancel one another out in most directions through destructive interference, they add constructively in a few specific directions, determined by Bragg's law:

$$2d \sin \theta = n\lambda$$

Here d is the spacing between diffracting planes, θ is the incident angle, n is any integer, and λ is the wavelength of the beam. These specific directions appear as spots on the diffraction pattern called *reflections*. Thus, X-ray diffraction results from an electromagnetic wave (the X-ray) impinging on a regular array of scatterers (the repeating arrangement of atoms within the crystal).

X-rays are used to produce the diffraction pattern because their wavelength λ is typically the same order of magnitude (1–100 angstroms) as the spacing d between planes in the crystal. In principle, any wave impinging on a regular array of scatterers produces diffraction, as predicted first by Francesco Maria Grimaldi in 1665. To produce significant diffraction, the spacing between the scatterers and the wavelength of the impinging wave should be similar in size. For illustration, the diffraction of sunlight through a bird's feather was first reported by James Gregory in the later 17th century. The first artificial diffraction gratings for visible light were constructed by David Rittenhouse in 1787, and Joseph von Fraunhofer in 1821. However, visible light has too long a wavelength (typically, 5500 angstroms) to observe diffraction from crystals. Prior to the first X-ray diffraction experiments, the spacings between lattice planes in a crystal were not known with certainty.

The idea that crystals could be used as a diffraction grating for X-rays arose in 1912 in a conversation between Paul Peter Ewald and Max von Laue in the English Garden in Munich. Ewald had proposed a resonator model of crystals for his thesis, but this model could not be validated using visible light, since the wavelength was much larger than the spacing between the resonators. Von Laue realised that electromagnetic radiation of a shorter wavelength was needed to observe such small spacings, and suggested that X-rays might have a wavelength comparable to the unit-cell spacing in crystals. Von Laue worked with two technicians, Walter Friedrich and his assistant Paul Knipping, to shine a beam of X-rays through a copper sulfate crystal and record its diffraction on a photographic plate. After being developed, the plate showed a large number of well-defined spots arranged in a pattern of intersecting circles around the spot produced by the central beam. Von Laue developed a law that connects the scattering angles and the size and orientation of the unit-cell spacings in the crystal, for which he was awarded the Nobel Prize in Physics in 1914.

Scattering

As described in the mathematical derivation below, the X-ray scattering is determined by the density of electrons within the crystal. Since the energy of an X-ray is much greater than that of a valence electron, the scattering may be modelled as Thomson scattering, the interaction of an electromagnetic ray with a free electron. This model is generally adopted to describe the polarization of the scattered radiation.

The intensity of Thomson scattering for one particle with mass m is:

$$I_o = I_e \left(\frac{q^4}{m^2 c^4} \right) \frac{1+\cos^2 2\theta}{2} = I_e 7.94.10^{-26} \frac{1+\cos^2 2\theta}{2} = I_e f$$

Hence the atomic nuclei, which are much heavier than an electron, contribute negligibly to the scattered X-rays.

Development from 1912 to 1920

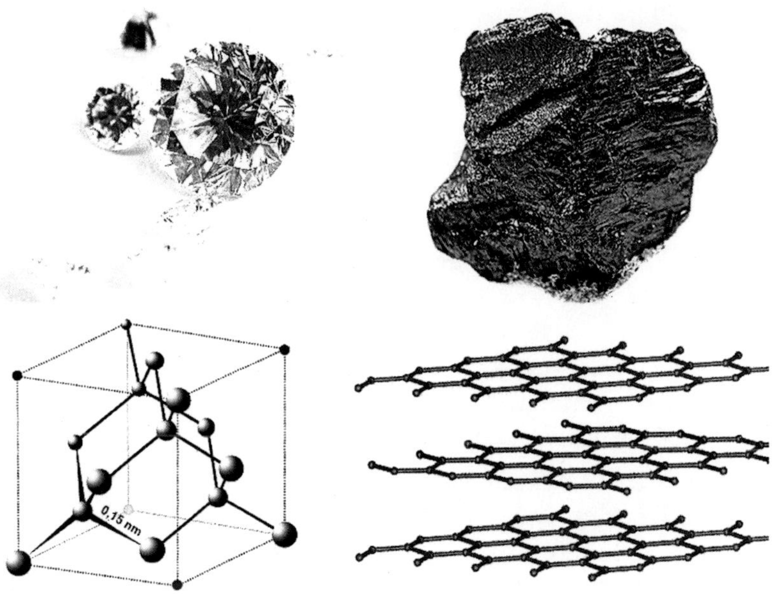

Figure: *Although diamonds (top left) and graphite (top right) are identical in chemical composition—being both pure carbon—X-ray crystallography revealed the arrangement of their atoms (bottom) accounts for their different properties. In diamond, the carbon atoms are arranged tetrahedrally and held together by single covalent bonds, making it strong in all directions. By contrast, graphite is composed of stacked sheets. Within the sheet, the bonding is covalent and has hexagonal symmetry, but there are no covalent bonds between the sheets, making graphite easy to cleave into flakes.*

After Von Laue's pioneering research, the field developed rapidly, most notably by physicists William Lawrence Bragg and his father William Henry Bragg. In 1912–1913, the younger Bragg developed Bragg's law, which connects the observed scattering with reflections from evenly spaced planes within the crystal. The Braggs, father and son, shared the 1915 Nobel Prize in Physics for their work in crystallography. The earliest structures were generally simple and marked by one-dimensional symmetry. However, as computational and experimental methods improved over the next decades, it became feasible to deduce reliable atomic positions for more complicated two- and three-dimensional arrangements of atoms in the unit-cell.

The potential of X-ray crystallography for determining the structure of molecules and minerals—then only known vaguely from chemical and hydrodynamic experiments—was realised immediately. The earliest structures were simple inorganic crystals and minerals, but even these revealed fundamental laws of physics and chemistry. The first atomic-resolution structure to be "solved" (i.e., determined) in 1914 was that of table salt. The distribution of electrons in the table-salt structure showed that crystals are not necessarily composed of covalently bonded molecules, and proved the existence of ionic compounds. The structure of diamond was solved in the same year, proving the tetrahedral arrangement of its chemical bonds and showing that the length of C–C single bond was 1.52 angstroms. Other early structures included copper, calcium fluoride (CaF_2, also known as *fluorite*), calcite ($CaCO_3$) and pyrite (FeS_2) in 1914; spinel ($MgAl_2O_4$) in 1915; the rutile and anatase forms of titanium dioxide (TiO_2) in 1916; pyrochroite $Mn(OH)_2$ and, by extension, brucite $Mg(OH)_2$ in 1919;. Also in 1919 sodium nitrate ($NaNO_3$) and caesium dichloroiodide ($CsICl_2$) were determined by Ralph Walter Graystone Wyckoff, and the wurtzite (hexagonal ZnS) structure became known in 1920.

The structure of graphite was solved in 1916 by the related method of powder diffraction, which was developed by Peter Debye and Paul Scherrer and, independently, by Albert Hull in 1917. The structure of graphite was determined from single-crystal diffraction in 1924 by two groups independently. Hull also used the powder method to determine the structures of various metals, such as iron and magnesium.

Cultural and Aesthetic Importance of X-Ray Crystallography

In what has been called his scientific autobiography, *The Development of X-ray Analysis*, Sir William Lawrence Bragg mentioned that he believed the field of crystallography was particularly welcoming to women because the techno-aesthetics of the molecular structures resembled textiles and household objects. Bragg was known to compare crystal formation to "curtains, wallpapers, mosaics, and roses."

In 1951, the Festival Pattern Group at the Festival of Britain hosted a collaborative group of textile manufacturers and experienced crystallographers to design lace and prints based on the X-ray crystallography of insulin, china clay, and hemoglobin. One of the leading scientists of the project was Dr. Helen Megaw (1907–2002), the Assistant Director of Research at the Cavendish Laboratory in Cambridge at the time. Megaw is credited as one of the central figures who took inspiration from crystal diagrams and saw their potential in design. In 2008, the Wellcome Collection in London curated an exhibition on the Festival Pattern Group called "From Atom to Patterns."

Contributions to Chemistry and Material Science

X-ray crystallography has led to a better understanding of chemical bonds and non-covalent interactions. The initial studies revealed the typical radii of atoms, and confirmed many theoretical models of chemical bonding, such as the tetrahedral bonding of carbon in the diamond structure, the octahedral bonding of metals observed in ammonium hexachloroplatinate (IV), and the resonance observed in the planar carbonate group and in aromatic molecules. Kathleen Lonsdale's 1928 structure of hexamethylbenzene established the hexagonal symmetry of benzene and showed a clear difference in bond length between the aliphatic C–C bonds and aromatic C–C bonds; this finding led to the idea of resonance between chemical bonds, which had profound consequences for the development of chemistry. Her conclusions were anticipated by William Henry Bragg, who published models of naphthalene and anthracene in 1921 based on other molecules, an early form of molecular replacement.

Also in the 1920s, Victor Moritz Goldschmidt and later Linus Pauling developed rules for eliminating chemically unlikely structures and for determining the relative sizes of atoms. These rules led to the structure of brookite (1928) and an understanding of the relative stability of the rutile, brookite and anatase forms of titanium dioxide.

The distance between two bonded atoms is a sensitive measure of the bond strength and its bond order; thus, X-ray crystallographic studies have led to the discovery of even more exotic types of bonding in inorganic chemistry, such as metal-metal double bonds, metal-metal quadruple bonds, and three-centre, two-electron bonds. X-ray crystallography—or, strictly speaking, an inelastic Compton scattering experiment—has also provided evidence for the partly covalent character of hydrogen bonds. In the field of organometallic chemistry, the X-ray structure of ferrocene initiated scientific studies of sandwich compounds, while that of Zeise's salt stimulated research into "back bonding" and metal-pi complexes. Finally, X-ray crystallography had a pioneering role in the development of supramolecular chemistry, particularly in clarifying the structures of the crown ethers and the principles of host-guest chemistry.

In material sciences, many complicated inorganic and organometallic systems have been analyzed using single-crystal methods, such as fullerenes, metalloporphyrins, and other complicated compounds. Single-crystal diffraction is also used in the pharmaceutical industry, due to recent problems with polymorphs. The major factors affecting the quality of single-crystal structures are the crystal's size and regularity; recrystallization is a commonly used technique to improve these factors in small-molecule crystals. The Cambridge Structural Database contains over 500,000 structures; over 99% of these structures were determined by X-ray diffraction.

Mineralogy and Metallurgy

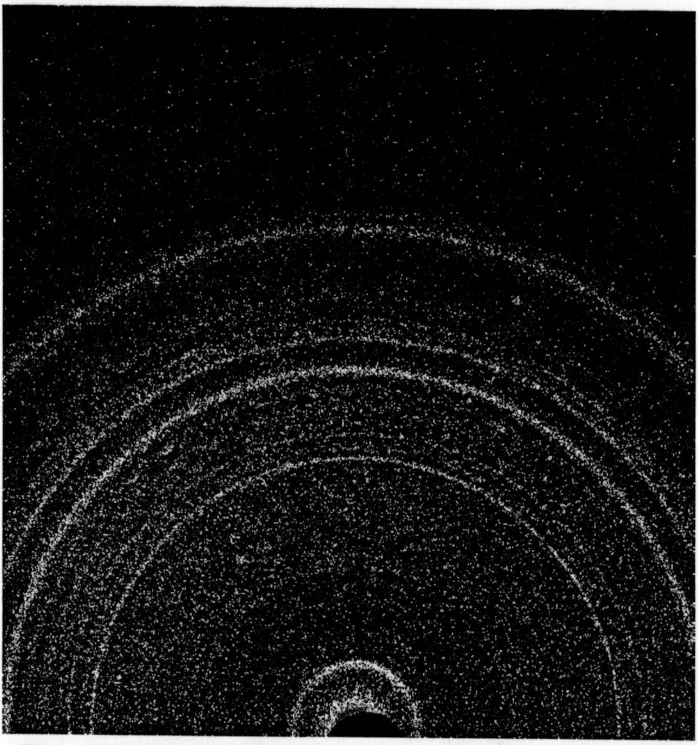

Figure: *First X-ray diffraction view of Martian soil - CheMin analysis reveals feldspar, pyroxenes, olivine and more (Curiosity rover at "Rocknest", October 17, 2012).*

Since the 1920s, X-ray diffraction has been the principal method for determining the arrangement of atoms in minerals and metals. The application of X-ray crystallography to mineralogy began with the structure of garnet, which was determined in 1924 by Menzer. A systematic X-ray crystallographic study of the silicates was undertaken in the 1920s. This study showed that, as the Si/O ratio is altered, the silicate crystals exhibit significant changes in their atomic arrangements. Machatschki extended these insights to minerals in which aluminium substitutes for the silicon atoms of the silicates. The first application of X-ray crystallography to metallurgy likewise occurred in the mid-1920s. Most notably, Linus Pauling's structure of the alloy Mg_2Sn led to his theory of the stability and structure of complex ionic crystals.

On October 17, 2012, the Curiosity rover on the planet Mars at "Rocknest" performed the first X-ray diffraction analysis of Martian soil. The results from the rover's CheMin analyzer revealed the presence of several minerals, including feldspar, pyroxenes and olivine, and suggested that the Martian soil in the sample was similar to the "weathered basaltic soils" of Hawaiian volcanoes.

Early Organic and Small Biological Molecules

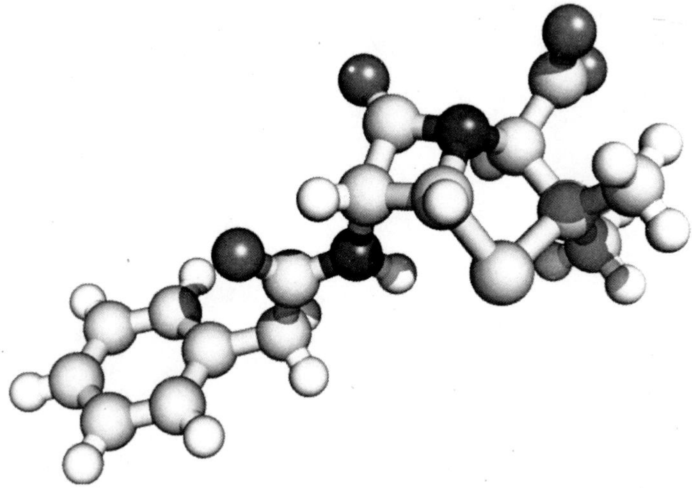

Figure: *The three-dimensional structure of penicillin, solved by Dorothy Crowfoot Hodgkin in 1945. The green, white, red, yellow and blue spheres represent atoms of carbon, hydrogen, oxygen, sulfur and nitrogen, respectively.*

The first structure of an organic compound, hexamethylenetetramine, was solved in 1923. This was followed by several studies of long-chain fatty acids, which are an important component of biological membranes. In the 1930s, the structures of much larger molecules with two-dimensional complexity began to be solved. A significant advance was the structure of phthalocyanine, a large planar molecule that is closely related to porphyrin molecules important in biology, such as heme, corrin and chlorophyll.

X-ray crystallography of biological molecules took off with Dorothy Crowfoot Hodgkin, who solved the structures of cholesterol (1937), penicillin (1946) and vitamin B12 (1956), for which she was awarded the Nobel Prize in Chemistry in 1964. In 1969, she succeeded in solving the structure of insulin, on which she worked for over thirty years.

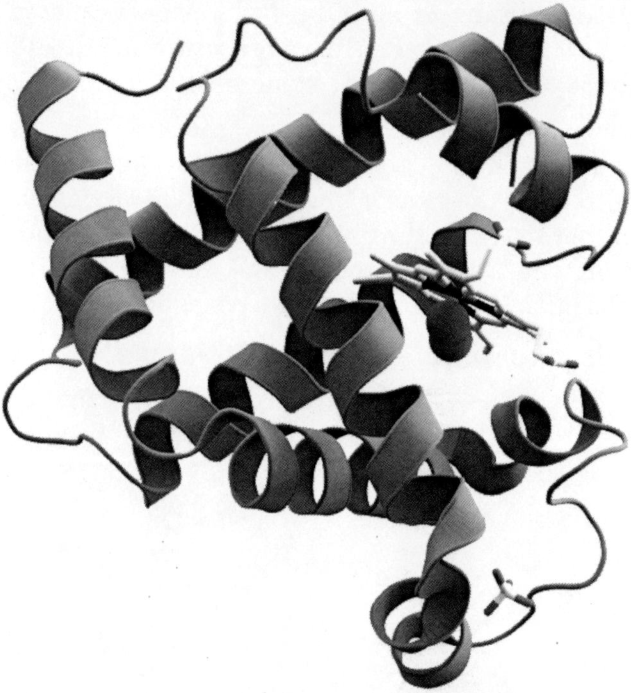

Figure: *Ribbon diagram of the structure of myoglobin, showing coloured alpha helices. Such proteins are long, linear molecules with thousands of atoms; yet the relative position of each atom has been determined with sub-atomic resolution by X-ray crystallography. Since it is difficult to visualize all the atoms at once, the ribbon shows the rough path of the protein polymer from its N-terminus (blue) to its C-terminus (red).*

Biological Macromolecular Crystallography

Crystal structures of proteins (which are irregular and hundreds of times larger than cholesterol) began to be solved in the late 1950s, beginning with the structure of sperm whale myoglobin by Sir John Cowdery Kendrew, for which he shared the Nobel Prize in Chemistry with Max Perutz in 1962. Since that success, over 86817 X-ray crystal structures of proteins, nucleic acids and other biological molecules have been determined. For comparison, the nearest competing method in terms of structures analyzed is nuclear magnetic resonance (NMR) spectroscopy, which has resolved 9561 chemical structures. Moreover, crystallography can solve structures of arbitrarily large molecules, whereas solution-state NMR is restricted to relatively small ones (less than 70 kDa). X-ray crystallography is now used routinely by scientists to determine how a pharmaceutical drug interacts with its protein target and what changes might improve it. However, intrinsic membrane proteins remain challenging to crystallize because they require detergents or other means to solubilize them in isolation, and such detergents often interfere with crystallization. Such membrane proteins are a large component of the genome and include many proteins of great physiological importance, such as ion channels and receptors.

Relationship to Other Scattering Techniques

Elastic vs. Inelastic Scattering

X-ray crystallography is a form of elastic scattering; the outgoing X-rays have the same energy, and thus same wavelength, as the incoming X-rays, only with altered direction. By contrast, *inelastic scattering* occurs when energy is transferred from the incoming X-ray to the crystal, e.g., by exciting an inner-shell electron to a higher energy level. Such inelastic scattering reduces the energy (or increases the wavelength) of the outgoing beam. Inelastic scattering is useful for probing such excitations of matter, but not in determining the distribution of scatterers within the matter, which is the goal of X-ray crystallography.

X-rays range in wavelength from 10 to 0.01 nanometres; a typical wavelength used for crystallography is 1 Å (0.1 nm), which is on the scale of covalent chemical bonds and the radius of a single atom. Longer-wavelength photons (such as ultraviolet radiation) would not have sufficient resolution to determine the atomic positions. At the other extreme, shorter-wavelength photons such as gamma rays are difficult to produce in large numbers, difficult to focus, and interact too strongly with matter, producing particle-antiparticle pairs. Therefore, X-rays are the "sweetspot" for wavelength when determining atomic-resolution structures from the scattering of electromagnetic radiation.

Other X-Ray Techniques

Other forms of elastic X-ray scattering include powder diffraction, SAXS and several types of X-ray fibre diffraction, which was used by Rosalind Franklin in determining the double-helix structure of DNA. In general, single-crystal X-ray diffraction offers more structural information than these other techniques; however, it requires a sufficiently large and regular crystal, which is not always available.

These scattering methods generally use *monochromatic* X-rays, which are restricted to a single wavelength with minor deviations. A broad spectrum of X-rays (that is, a blend of X-rays with different wavelengths) can also be used to carry out X-ray diffraction, a technique known as the Laue method. This is the method used in the original discovery of X-ray diffraction. Laue scattering provides much structural information with only a short exposure to the X-ray beam, and is therefore used in structural studies of very rapid events (Time resolved crystallography). However, it is not as well-suited as monochromatic scattering for determining the full atomic structure of a crystal and therefore works better with crystals with relatively simple atomic arrangements.

The Laue back reflection mode records X-rays scattered backwards from a broad spectrum source. This is useful if the sample is too thick for X-rays to transmit through it. The diffracting planes in the crystal are determined by knowing that the normal to the diffracting plane bisects the angle between the incident beam and the diffracted beam. A Greninger chart can be used to interpret the back reflection Laue photograph.

Electron and Neutron Diffraction

Other particles, such as electrons and neutrons, may be used to produce a diffraction pattern. Although electron, neutron, and X-ray scattering are based on different physical processes, the resulting diffraction patterns are analyzed using the same coherent diffraction imaging techniques.

As derived below, the electron density within the crystal and the diffraction patterns are related by a simple mathematical method, the Fourier transform, which allows the density to be calculated relatively easily from the patterns. However, this works only if the scattering is *weak*, i.e., if the scattered beams are much less intense than the incoming beam. Weakly scattered beams pass through the remainder of the crystal without undergoing a second scattering event. Such re-scattered waves are called "secondary scattering" and hinder the analysis. Any sufficiently thick crystal will produce secondary scattering, but since X-rays interact relatively weakly with the electrons, this is generally not a significant concern. By contrast, electron beams may produce strong secondary scattering even for relatively thin crystals (>100 nm). Since this thickness corresponds to the diameter of many viruses, a promising direction is the electron diffraction of isolated macromolecular assemblies, such as viral capsids and molecular machines, which may be carried out with a cryo-electron microscope. Moreover the strong interaction of electrons with matter (about 1000 times stronger than for X-rays) allows determination of the atomic structure of extremely small volumes. The field of applications for electron crystallography ranges from bio molecules like membrane proteins over organic thin films to the complex structures of (nanocrystalline) intermetallic compounds and zeolites.

Neutron diffraction is an excellent method for structure determination, although it has been difficult to obtain intense, monochromatic beams of neutrons in sufficient quantities. Traditionally, nuclear reactors have been used, although the new Spallation Neutron Source holds much promise in the near future. Being uncharged, neutrons scatter much more readily from the atomic nuclei rather than from the electrons. Therefore, neutron scattering is very useful for observing the positions of light atoms with few electrons, especially hydrogen, which is essentially invisible in the X-ray diffraction. Neutron scattering also has the remarkable property that the solvent can be made invisible by adjusting the ratio of normal water, H_2O, and heavy water, D_2O.

Synchrotron Radiation

Synchrotron Radiation is one of the brightest lights on earth. It is the single most powerful tool available to X-ray crystallographers. It is made of X-ray beams generated in large machines called synchrotrons. These machines accelerate electrically charged particles, often electrons, to nearly the speed of light, then whip them around a huge, hollow metal ring.

Synchrotrons were originally designed for use by high-energy physicists studying subatomic particles and cosmic phenomena. The largest component of each synchrotron is its electron storage ring. This ring is actually not a perfect circle, but a many-sided polygon. At each corner of the polygon, precisely aligned magnets bend the electron stream, forcing it to stay in the ring. Each time the electrons' path is bent, they emit bursts of energy in the form of electromagnetic radiation. Because particles in a synchrotron are hurtling at nearly the speed of light, they emit intense radiation, including lots of high-energy X-rays.

Methods

Overview of Single-Crystal X-Ray Diffraction

The oldest and most precise method of X-ray crystallography is *single-crystal X-ray diffraction*, in which a beam of X-rays strikes a single crystal, producing scattered beams. When they land on a piece of film or other detector, these beams make a *diffraction pattern* of spots; the strengths and angles of these beams are recorded as the crystal is gradually rotated. Each spot is called a *reflection*, since it corresponds to the reflection of the X-rays from one set of evenly spaced planes within the crystal. For single crystals of sufficient purity and regularity, X-ray diffraction data can determine the mean chemical bond lengths and angles to within a few thousandths of an angstrom and to within a few tenths of a degree, respectively. The atoms in a crystal are not static, but oscillate about their mean positions, usually by less than a few tenths of an angstrom. X-ray crystallography allows measuring the size of these oscillations.

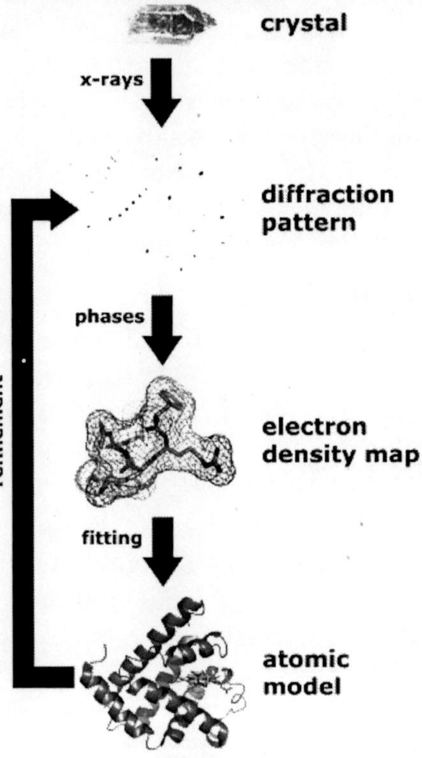

***Figure:** Workflow for solving the structure of a molecule by X-ray crystallography.*

Procedure

The technique of single-crystal X-ray crystallography has three basic steps. The first—and often most difficult—step is to obtain an adequate crystal of the material under study. The crystal should be sufficiently large (typically larger than 0.1 mm in all dimensions), pure in composition and regular in structure, with no significant internal imperfections such as cracks or twinning.

In the second step, the crystal is placed in an intense beam of X-rays, usually of a single wavelength (*monochromatic X-rays*), producing the regular pattern of reflections. As the crystal is gradually rotated, previous reflections disappear and new ones appear; the intensity of every spot is recorded at every orientation of the crystal. Multiple data sets may have to be collected, with each set covering slightly more than half a full rotation of the crystal and typically containing tens of thousands of reflections.

In the third step, these data are combined computationally with complementary chemical information to produce and refine a model of the arrangement of atoms within the crystal. The final, refined model of the atomic arrangement—now called a *crystal structure*—is usually stored in a public database.

Limitations

As the crystal's repeating unit, its unit cell, becomes larger and more complex, the atomic-level picture provided by X-ray crystallography becomes less well-resolved (more "fuzzy") for a given number of observed reflections. Two limiting cases of X-ray crystallography—"small-molecule" and "macromolecular" crystallography—are often discerned. *Small-molecule crystallography* typically involves crystals with fewer than 100 atoms in their asymmetric unit; such crystal structures are usually so well resolved that the atoms can be discerned as isolated "blobs" of electron density. By contrast, *macromolecular crystallography* often involves tens of thousands of atoms in the unit cell. Such crystal structures are generally less well-resolved (more "smeared out"); the atoms and chemical bonds appear as tubes of electron density, rather than as isolated atoms. In general, small molecules are also easier to crystallize than macromolecules; however, X-ray crystallography has proven possible even for viruses with hundreds of thousands of atoms.

Crystallization

Figure: *A protein crystal seen under a microscope. Crystals used in X-ray crystallography may be smaller than a millimetre across.*

Although crystallography can be used to characterize the disorder in an impure or irregular crystal, crystallography generally requires a pure crystal of high regularity to solve the structure of a complicated arrangement of atoms. Pure, regular crystals can sometimes be obtained from natural or synthetic materials, such as samples of metals, minerals or other macroscopic materials. The regularity of such crystals can sometimes be improved with macromolecular crystal annealing and other methods. However, in many cases, obtaining a diffraction-quality crystal is the chief barrier to solving its atomic-resolution structure.

Small-molecule and macromolecular crystallography differ in the range of possible techniques used to produce diffraction-quality crystals. Small molecules generally have few degrees of conformational freedom, and may be crystallized by a wide range of methods, such as chemical vapor deposition and recrystallization. By contrast, macromolecules generally have many degrees of freedom and their crystallization must be carried out to maintain a stable structure. For example, proteins and larger RNA molecules cannot be crystallized if their tertiary structure has been unfolded; therefore, the range of crystallization conditions is restricted to solution conditions in which such molecules remain folded.

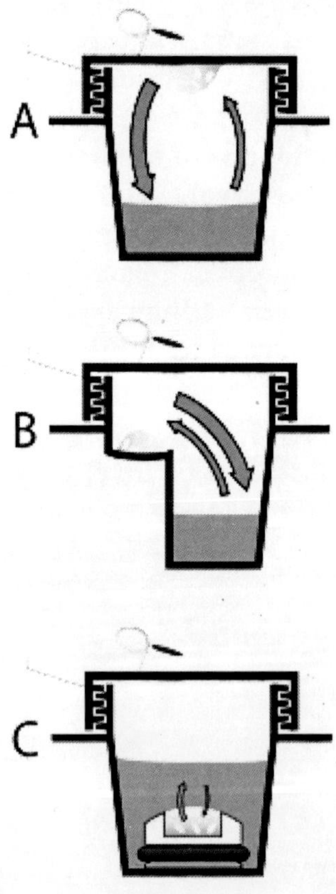

***Figure:** Three methods of preparing crystals,*
A: Hanging drop. B: Sitting drop. C: Microdialysis

Protein crystals are almost always grown in solution. The most common approach is to lower the solubility of its component molecules very gradually; if this is done too quickly, the molecules will precipitate from solution, forming a useless dust or amorphous gel on the bottom of the container. Crystal growth in solution is characterized by two steps: *nucleation* of a microscopic crystallite (possibly having only 100 molecules), followed by *growth* of that crystallite, ideally to a diffraction-quality crystal. The solution conditions that favour the first step (nucleation) are not always the same conditions that favour the second step (subsequent growth). The crystallographer's goal is to identify solution conditions that favour the development of a single, large crystal, since larger crystals offer improved resolution of the molecule. Consequently, the solution conditions should *disfavour* the first step (nucleation) but *favour* the second (growth), so that only one large crystal forms per droplet. If nucleation is favoured too much, a shower of small crystallites will form in the droplet, rather than one large crystal; if favoured too little, no crystal will form whatsoever.

It is extremely difficult to predict good conditions for nucleation or growth of well-ordered crystals. In practice, favourable conditions are identified by *screening*; a very large batch of the molecules is prepared, and a wide variety of crystallization solutions are tested. Hundreds, even thousands, of solution conditions are generally tried before finding the successful one. The various conditions can use one or more physical mechanisms to lower the solubility of the molecule; for example, some may change the pH, some contain salts of the Hofmeister series or chemicals that lower the dielectric constant of the solution, and still others contain large polymers such as polyethylene glycol that drive the molecule out of solution by entropic effects. It is also common to try several temperatures for encouraging crystallization, or to gradually lower the temperature so that the solution becomes supersaturated. These methods require large amounts of the target molecule, as they use high concentration of the molecule(s) to be crystallized. Due to the difficulty in obtaining such large quantities (milligrams) of crystallization grade protein, robots have been developed that are capable of accurately dispensing crystallization trial drops that are in the order of 100 nanoliters in volume. This means that 10-fold less protein is used per-experiment when compared to crystallization trials setup by hand (in the order of 1 microliter).

Several factors are known to inhibit or mar crystallization. The growing crystals are generally held at a constant temperature and protected from shocks or vibrations that might disturb their crystallization. Impurities in the molecules or in the crystallization solutions are often inimical to crystallization. Conformational flexibility in the molecule also tends to make crystallization less likely, due to entropy. Ironically, molecules that tend to self-assemble into regular helices are often unwilling to assemble into crystals. Crystals can be marred by twinning, which can occur when a unit cell can pack equally favourably in multiple orientations; although recent advances in computational methods may allow solving the structure of some twinned crystals. Having failed to crystallize a target molecule, a crystallographer may try again with a slightly modified version of the molecule; even small changes in molecular properties can lead to large differences in crystallization behavior.

Data Collection

Mounting the Crystal

Figure: *Animation showing the five motions possible with a four-circle kappa goniometre. The rotations about each of the four angles φ, κ, ω and 2θ leave the crystal within the X-ray beam, but change the crystal orientation. The detector (red box) can be slid closer or further away from the crystal, allowing higher resolution data to be taken (if closer) or better discernment of the Bragg peaks (if further away).*

The crystal is mounted for measurements so that it may be held in the X-ray beam and rotated. There are several methods of mounting. In the past, crystals were loaded into glass capillaries with the crystallization solution (the mother liquor). Nowadays, crystals of small molecules are a typically attached with oil or glue to a glass fibre or a loop, which is made of nylon or plastic and attached to a solid rod. Protein crystals are scooped up by a loop, then flash-frozen with liquid nitrogen. This freezing reduces the radiation damage of the X-rays, as well as the noise in the Bragg peaks due to thermal motion (the Debye-Waller effect). However, untreated protein crystals often crack if flash-frozen; therefore, they are generally pre-soaked in a cryoprotectant solution before freezing. Unfortunately, this pre-soak may itself cause the crystal to crack, ruining it for crystallography. Generally, successful cryo-conditions are identified by trial and error.

The capillary or loop is mounted on a goniometre, which allows it to be positioned accurately within the X-ray beam and rotated. Since both the crystal and the beam are often very small, the crystal must be centred within the beam to within ~25 micrometres accuracy, which is aided by a camera focused on the crystal. The most common type of goniometre is the "kappa goniometre", which offers three angles of rotation: the ω angle, which rotates about an axis perpendicular to the beam; the κ angle, about an axis at ~50° to the ω axis; and, finally, the φ angle about the loop/capillary axis. When the κ angle is zero, the ω and φ axes are aligned. The κ rotation allows for convenient mounting of the crystal, since the arm in which the crystal is mounted may be swung out towards the crystallographer. The oscillations carried out during data collection (mentioned below) involve the ω axis only. An older type of goniometre is the four-circle goniometre, and its relatives such as the six-circle goniometre.

X-Ray Sources

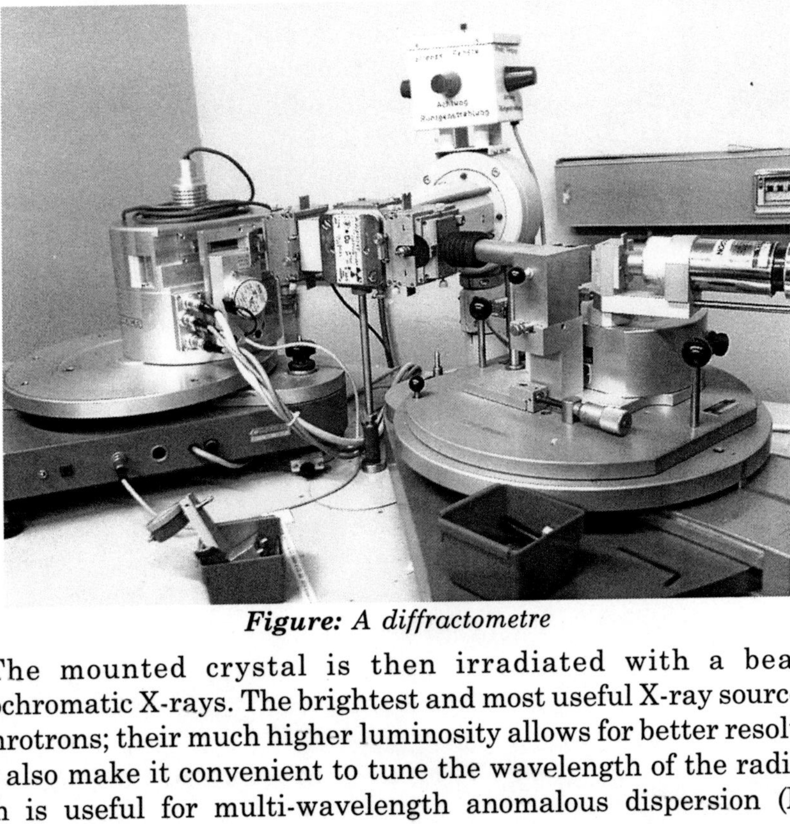

***Figure:** A diffractometre*

The mounted crystal is then irradiated with a beam of monochromatic X-rays. The brightest and most useful X-ray sources are synchrotrons; their much higher luminosity allows for better resolution. They also make it convenient to tune the wavelength of the radiation, which is useful for multi-wavelength anomalous dispersion (MAD) phasing, described below. Synchrotrons are generally national facilities, each with several dedicated beamlines where data is collected around the clock, seven days a week.

Smaller, X-ray generators are often used in laboratories to check the quality of crystals before bringing them to a synchrotron and sometimes to solve a crystal structure. In such systems, electrons are boiled off of a cathode and accelerated through a strong electric potential of ~50 kV; having reached a high speed, the electrons collide with a metal plate, emitting *bremsstrahlung* and some strong spectral lines corresponding to the excitation of inner-shell electrons of the metal. The most common metal used is copper, which can be kept cool easily, due to its high thermal conductivity, and which produces strong K_α and K_β lines. The K_β line is sometimes suppressed with a thin (~10 μm) nickel foil. The simplest and cheapest variety of sealed X-ray tube has a stationary anode (the Crookes tube) and run with ~2 kW of electron beam power. The more expensive variety has a rotating-anode type source that run with ~14 kW of e-beam power.

X-rays are generally filtered (by use of X-Ray Filters) to a single wavelength (made monochromatic) and collimated to a single direction before they are allowed to strike the crystal. The filtering not only simplifies the data analysis, but also removes radiation that degrades the crystal without contributing useful information. Collimation is done either with a collimator (basically, a long tube) or with a clever arrangement of gently curved mirrors. Mirror systems are preferred for small crystals (under 0.3 mm) or with large unit cells (over 150 Å)

Recording the Reflections

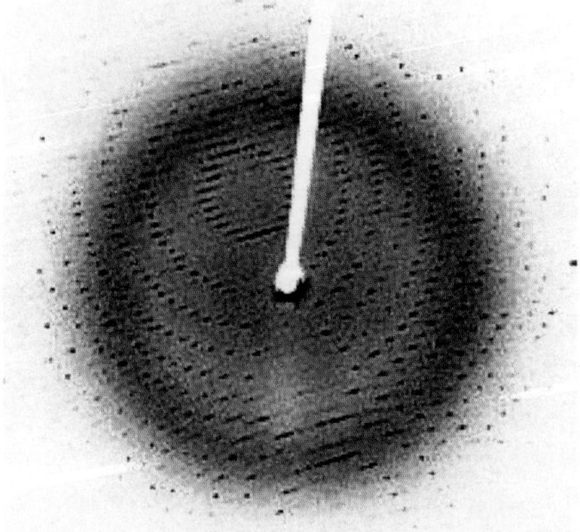

Figure: *An X-ray diffraction pattern of a crystallized enzyme. The pattern of spots* (reflections) *and the relative strength of each spot* (intensities) *can be used to determine the structure of the enzyme.*

When a crystal is mounted and exposed to an intense beam of X-rays, it scatters the X-rays into a pattern of spots or *reflections* that can be observed on a screen behind the crystal. A similar pattern may be seen by shining a laser pointer at a compact disc. The relative intensities of these spots provide the information to determine the arrangement of molecules within the crystal in atomic detail. The intensities of these reflections may be recorded with photographic film, an area detector or with a charge-coupled device (CCD) image sensor. The peaks at small angles correspond to low-resolution data, whereas those at high angles represent high-resolution data; thus, an upper limit on the eventual resolution of the structure can be determined from the first few images. Some measures of diffraction quality can be determined at this point, such as the mosaicity of the crystal and its overall disorder, as observed in the peak widths. Some pathologies of the crystal that would render it unfit for solving the structure can also be diagnosed quickly at this point.

One image of spots is insufficient to reconstruct the whole crystal; it represents only a small slice of the full Fourier transform. To collect all the necessary information, the crystal must be rotated step-by-step through 180°, with an image recorded at every step; actually, slightly more than 180° is required to cover reciprocal space, due to the curvature of the Ewald sphere. However, if the crystal has a higher symmetry, a smaller angular range such as 90° or 45° may be recorded. The rotation axis should be changed at least once, to avoid developing a "blind spot" in reciprocal space close to the rotation axis. It is customary to rock the crystal slightly (by 0.5–2°) to catch a broader region of reciprocal space.

Multiple data sets may be necessary for certain phasing methods. For example, MAD phasing requires that the scattering be recorded at least three (and usually four, for redundancy) wavelengths of the incoming X-ray radiation. A single crystal may degrade too much during the collection of one data set, owing to radiation damage; in such cases, data sets on multiple crystals must be taken.

Data Analysis

Crystal Symmetry, Unit Cell, and Image Scaling

The recorded series of two-dimensional diffraction patterns, each corresponding to a different crystal orientation, is converted into a three-dimensional model of the electron density; the conversion uses the mathematical technique of Fourier transforms, which is explained below. Each spot corresponds to a different type of variation in the electron density; the crystallographer must determine *which* variation corresponds to *which* spot (*indexing*), the relative strengths of the spots in different images (*merging and scaling*) and how the variations should be combined to yield the total electron density (*phasing*).

Data processing begins with *indexing* the reflections. This means identifying the dimensions of the unit cell and which image peak corresponds to which position in reciprocal space. A byproduct of indexing is to determine the symmetry of the crystal, i.e., its *space group*. Some space groups can be eliminated from the beginning. For example, reflection symmetries cannot be observed in chiral molecules; thus, only 65 space groups of 230 possible are allowed for protein molecules which are almost always chiral. Indexing is generally accomplished using an *autoindexing* routine. Having assigned symmetry, the data is then *integrated*. This converts the hundreds of images containing the thousands of reflections into a single file, consisting of (at the very least) records of the Miller index of each reflection, and an intensity for each reflection (at this state the file often also includes error estimates and measures of partiality (what part of a given reflection was recorded on that image)).

A full data set may consist of hundreds of separate images taken at different orientations of the crystal. The first step is to merge and scale these various images, that is, to identify which peaks appear in two or more images (*merging*) and to scale the relative images so that they have a consistent intensity scale. Optimizing the intensity scale is critical because the relative intensity of the peaks is the key information from which the structure is determined. The repetitive technique of crystallographic data collection and the often high symmetry of crystalline materials cause the diffractometre to record many symmetry-equivalent reflections multiple times. This allows calculating the symmetry-related R-factor, a reliability index based upon how similar are the measured intensities of symmetry-equivalent reflections, thus assessing the quality of the data.

Initial Phasing

The data collected from a diffraction experiment is a reciprocal space representation of the crystal lattice. The position of each diffraction 'spot' is governed by the size and shape of the unit cell, and the inherent symmetry within the crystal. The intensity of each diffraction 'spot' is recorded, and this intensity is proportional to the square of the *structure factor* amplitude. The structure factor is a complex number containing information relating to both the amplitude and phase of a wave. In order to obtain an interpretable *electron density map*, both amplitude and phase must be known (an electron density map allows a crystallographer to build a starting model of the molecule). The phase cannot be directly recorded during a diffraction experiment: this is known as the phase problem. Initial phase estimates can be obtained in a variety of ways:

- *Ab initio* phasing or direct methods – This is usually the method of choice for small molecules (<1000 non-hydrogen atoms), and has been used successfully to solve the phase problems for small proteins. If the resolution of the data is better than 1.4 Å (140 pm), direct methods can be used to obtain phase information, by exploiting known phase relationships between certain groups of reflections.
- Molecular replacement – if a related structure is known, it can be used as a search model in molecular replacement to determine the orientation and position of the molecules within the unit cell. The phases obtained this way can be used to generate *electron density maps.*
- Anomalous X-ray scattering (*MAD or SAD phasing*) – the X-ray wavelength may be scanned past an absorption edge of an atom, which changes the scattering in a known way. By recording full sets of reflections at three different wavelengths (far below, far above and in the middle of the absorption edge) one can solve for the substructure of the anomalously diffracting atoms and thence the structure of the whole molecule. The most popular method of incorporating anomalous scattering atoms into proteins is to express the protein in a methionine auxotroph (a host incapable of synthesizing methionine) in a media rich in seleno-methionine, which contains selenium atoms. A MAD experiment can then be conducted around the absorption edge, which should then yield the position of any methionine residues within the protein, providing initial phases.
- Heavy atom methods (multiple isomorphous replacement) – If electron-dense metal atoms can be introduced into the crystal, direct methods or Patterson-space methods can be used to determine their location and to obtain initial phases. Such heavy atoms can be introduced either by soaking the crystal in a heavy atom-containing solution, or by co-crystallization (growing the crystals in the presence of a heavy atom). As in MAD phasing, the changes in the scattering amplitudes can be interpreted to yield the phases. Although this is the original method by which protein crystal structures were solved, it has largely been superseded by MAD phasing with selenomethionine.

Model Building and Phase Refinement

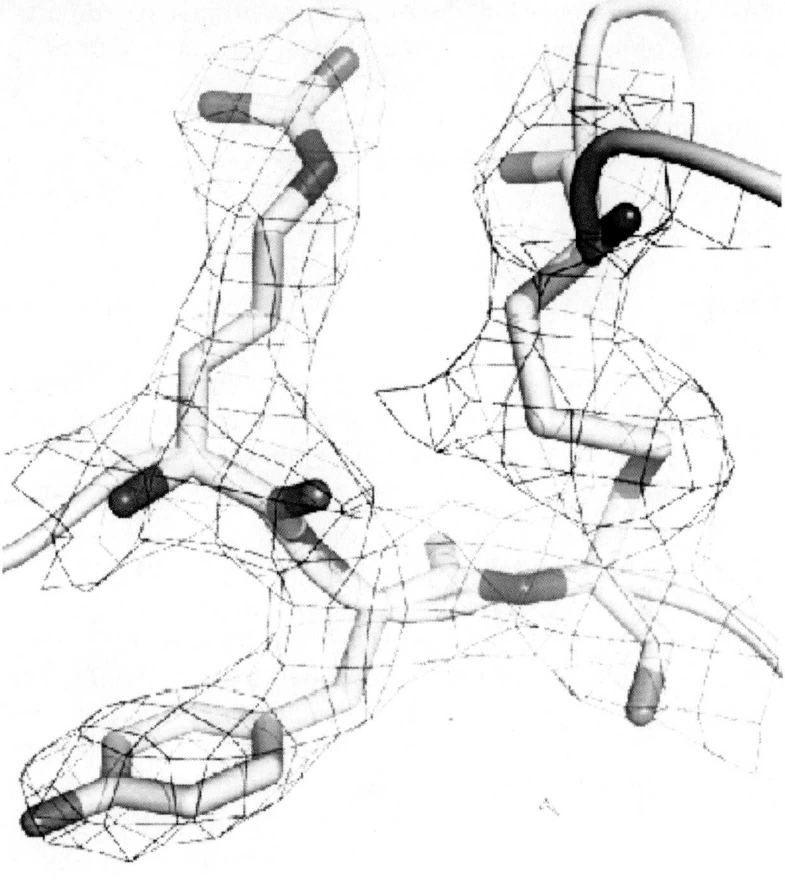

Figure: *A protein crystal structure at 2.7 Å resolution. The mesh encloses the region in which the electron density exceeds a given threshold. The straight segments represent chemical bonds between the non-hydrogen atoms of an arginine (upper left), a tyrosine (lower left), a disulfide bond (upper right, in yellow), and some peptide groups (running left-right in the middle). The two curved green tubes represent spline fits to the polypeptide backbone.*

Having obtained initial phases, an initial model can be built. This model can be used to refine the phases, leading to an improved model, and so on. Given a model of some atomic positions, these positions and their respective Debye-Waller factors (or B-factors, accounting for the thermal motion of the atom) can be refined to fit the observed diffraction data, ideally yielding a better set of phases. A new model can then be fit to the new electron density map and a further round of refinement is carried out. This continues until the correlation between the diffraction data and the model is maximized. The agreement is measured by an R-factor defined as

$$R = \frac{\sum_{\text{all reflections}} |F_o - F_c|}{\sum_{\text{all reflections}} |F_o|}$$

where F is the structure factor. A similar quality criterion is R_{free}, which is calculated from a subset (~10%) of reflections that were not included in the structure refinement. Both R factors depend on the resolution of the data. As a rule of thumb, R_{free} should be approximately the resolution in angstroms divided by 10; thus, a data-set with 2 Å resolution should yield a final $R_{\text{free}} \sim 0.2$. Chemical bonding features such as stereochemistry, hydrogen bonding and distribution of bond lengths and angles are complementary measures of the model quality. Phase bias is a serious problem in such iterative model building. *Omit maps* are a common technique used to check for this.

It may not be possible to observe every atom of the crystallized molecule – it must be remembered that the resulting electron density is an average of all the molecules within the crystal. In some cases, there is too much residual disorder in those atoms, and the resulting electron density for atoms existing in many conformations is smeared to such an extent that it is no longer detectable in the electron density map. Weakly scattering atoms such as hydrogen are routinely invisible. It is also possible for a single atom to appear multiple times in an electron density map, e.g., if a protein sidechain has multiple (<4) allowed conformations. In still other cases, the crystallographer may detect that the covalent structure deduced for the molecule was incorrect, or changed. For example, proteins may be cleaved or undergo post-translational modifications that were not detected prior to the crystallization.

Deposition of the Structure

Once the model of a molecule's structure has been finalized, it is often deposited in a crystallographic database such as the Cambridge Structural Database (for small molecules), the Inorganic Crystal Structure Database (ICSD) (for inorganic compounds) or the Protein Data Bank (for protein structures). Many structures obtained in private commercial ventures to crystallize medicinally relevant proteins are not deposited in public crystallographic databases.

Diffraction Theory

The main goal of X-ray crystallography is to determine the density of electrons *f*(r) throughout the crystal, where r represents the three-dimensional position vector within the crystal. To do this, X-ray scattering is used to collect data about its Fourier transform $F(\mathrm{q})$, which is inverted mathematically to obtain the density defined in real space, using the formula

$$f(\mathrm{r}) = \frac{1}{(2\pi)^3} \int F(\mathrm{q}) e^{i\mathrm{q}\cdot\mathrm{r}} d\mathrm{q}$$

where the integral is taken over all values of q. The three-dimensional real vector q represents a point in reciprocal space, that is, to a particular oscillation in the electron density as one moves in the direction in which q points.

The length of q corresponds to 2π divided by the wavelength of the oscillation. The corresponding formula for a Fourier transform will be used below

$$F(\mathrm{q}) = \int f(\mathrm{r}) e^{-i\mathrm{q}\cdot\mathrm{r}} d\mathrm{r}$$

where the integral is summed over all possible values of the position vector r within the crystal.

The Fourier transform $F(\mathrm{q})$ is generally a complex number, and therefore has a magnitude $|F(\mathrm{q})|$ and a phase $\varphi(\mathrm{q})$ related by the equation

$$F(\mathrm{q}) = |F(\mathrm{q})| e^{i\phi(\mathrm{q})}$$

The intensities of the reflections observed in X-ray diffraction give us the magnitudes $|F(\mathrm{q})|$ but not the phases $\varphi(\mathrm{q})$. To obtain the phases, full sets of reflections are collected with known alterations to the scattering, either by modulating the wavelength past a certain absorption edge or by adding strongly scattering (i.e., electron-dense) metal atoms such as mercury. Combining the magnitudes and phases yields the full Fourier transform $F(\mathrm{q})$, which may be inverted to obtain the electron density $f(\mathrm{r})$.

Crystals are often idealized as being *perfectly* periodic. In that ideal case, the atoms are positioned on a perfect lattice, the electron density is perfectly periodic, and the Fourier transform $F(\mathrm{q})$ is zero except when q belongs to the reciprocal lattice (the so-called *Bragg peaks*). In reality, however, crystals are not perfectly periodic; atoms vibrate about their mean position, and there may be disorder of various types, such as mosaicity, dislocations, various point defects, and heterogeneity in the conformation of crystallized molecules.

Therefore, the Bragg peaks have a finite width and there may be significant *diffuse scattering*, a continuum of scattered X-rays that fall between the Bragg peaks.

Intuitive Understanding by Bragg's Law

An intuitive understanding of X-ray diffraction can be obtained from the Bragg model of diffraction. In this model, a given reflection is associated with a set of evenly spaced sheets running through the crystal, usually passing through the centres of the atoms of the crystal lattice. The orientation of a particular set of sheets is identified by its three Miller indices (h, k, l), and let their spacing be noted by d. William Lawrence Bragg proposed a model in which the incoming X-rays are scattered specularly (mirror-like) from each plane; from that assumption, X-rays scattered from adjacent planes will combine constructively (constructive interference) when the angle between the plane and the X-ray results in a path-length difference that is an integer multiple n of the X-ray wavelength λ.

$$2d \sin\theta = n\lambda$$

A reflection is said to be *indexed* when its Miller indices (or, more correctly, its reciprocal lattice vector components) have been identified from the known wavelength and the scattering angle 2θ. Such indexing gives the unit-cell parameters, the lengths and angles of the unit-cell, as well as its space group. Since Bragg's law does not interpret the relative intensities of the reflections, however, it is generally inadequate to solve for the arrangement of atoms within the unit-cell; for that, a Fourier transform method must be carried out.

Scattering as a Fourier Transform

The incoming X-ray beam has a polarization and should be represented as a vector wave; however, for simplicity, let it be represented here as a scalar wave. We also ignore the complication of the time dependence of the wave and just focus on the wave's spatial dependence. Plane waves can be represented by a wave vector k_{in}, and so the strength of the incoming wave at time $t=0$ is given by

$$Ae^{ik_{in}\cdot r}$$

At position r within the sample, let there be a density of scatterers $f(r)$; these scatterers should produce a scattered spherical wave of amplitude proportional to the local amplitude of the incoming wave times the number of scatterers in a small volume dV about r

$$\text{amplitude of scattered wave} = Ae^{ik\cdot r} Sf(r)dV$$

where S is the proportionality constant.

Let's consider the fraction of scattered waves that leave with an outgoing wave-vector of k_{out} and strike the screen at r_{screen}. Since no energy is lost (elastic, not inelastic scattering), the wavelengths are the same as are the magnitudes of the wave-vectors $|k_{in}| = |k_{out}|$. From the time that the photon is scattered at r until it is absorbed at r_{screen}, the photon undergoes a change in phase

$$e^{ik_{out}\cdot(r_{screen}-r)}$$

The net radiation arriving at r_{screen} is the sum of all the scattered waves throughout the crystal

$$AS\int dr f(r) e^{ik_{in}\cdot r} e^{ik_{out}\cdot(r_{screen}-r)} = ASe^{ik_{out}\cdot r_{screen}} \int dr f(r) e^{i(k_{in}-k_{out})\cdot r}$$

which may be written as a Fourier transform

$$ASe^{ik_{out}\cdot r_{screen}} \int dr f(r) e^{-iq\cdot r} = ASe^{ik_{out}\cdot r_{screen}} F(q)$$

where $q = k_{out} - k_{in}$. The measured intensity of the reflection will be square of this amplitude

$$A^2 S^2 |F(q)|^2$$

Friedel and Bijvoet Mates

For every reflection corresponding to a point q in the reciprocal space, there is another reflection of the same intensity at the opposite point -q. This opposite reflection is known as the *Friedel mate* of the original reflection.

This symmetry results from the mathematical fact that the density of electrons $f(r)$ at a position r is always a real number. As noted above, $f(r)$ is the inverse transform of its Fourier transform $F(q)$; however, such an inverse transform is a complex number in general.

To ensure that $f(r)$ is real, the Fourier transform $F(q)$ must be such that the Friedel mates $F(-q)$ and $F(q)$ are complex conjugates of one another. Thus, $F(-q)$ has the same magnitude as $F(q)$ but they have the opposite phase, i.e., $\varphi(q) = -\varphi(q)$

$$F(-q) = |F(-q)| e^{i\phi(-q)} = F^*(q) = |F(q)| e^{-i\phi(q)}$$

The equality of their magnitudes ensures that the Friedel mates have the same intensity $|F|^2$. This symmetry allows one to measure the full Fourier transform from only half the reciprocal space, e.g., by rotating the crystal slightly more than 180° instead of a full 360° revolution.

In crystals with significant symmetry, even more reflections may have the same intensity (Bijvoet mates); in such cases, even less of the reciprocal space may need to be measured.

In favourable cases of high symmetry, sometimes only 90° or even only 45° of data are required to completely explore the reciprocal space.

The Friedel-mate constraint can be derived from the definition of the inverse Fourier transform

$$f(\mathrm{r}) = \int \frac{d\mathrm{q}}{(2\pi)^3} F(\mathrm{q}) e^{i\mathrm{q}\cdot\mathrm{r}} = \int \frac{d\mathrm{q}}{(2\pi)^3} |F(\mathrm{q})| e^{i\phi(\mathrm{q})} e^{i\mathrm{q}\cdot\mathrm{r}}.$$

Since Euler's formula states that $e^{ix} = \cos(x) + i\sin(x)$, the inverse Fourier transform can be separated into a sum of a purely real part and a purely imaginary part

$$f(\mathrm{r}) = \int \frac{d\mathrm{q}}{(2\pi)^3} |F(\mathrm{q})| e^{i(\phi+\mathrm{q}\cdot\mathrm{r})} = \int \frac{d\mathrm{q}}{(2\pi)^3} |F(\mathrm{q})| \cos(\phi + \mathrm{q}\cdot\mathrm{r}) + i\int \frac{d\mathrm{q}}{(2\pi)^3} |F(\mathrm{q})| \sin(\phi + \mathrm{q}\cdot\mathrm{r}) = I_{\cos} + iI_{\sin}$$

The function $f(\mathrm{r})$ is real if and only if the second integral $I_{\sin}$ is zero for all values of r. In turn, this is true if and only if the above constraint is satisfied

$$I_{\sin} = \int \frac{d\mathrm{q}}{(2\pi)^3} |F(\mathrm{q})| \sin(\phi + \mathrm{q}\cdot\mathrm{r}) = \int \frac{d\mathrm{q}}{(2\pi)^3} |F(-\mathrm{q})| \sin(-\phi - \mathrm{q}\cdot\mathrm{r}) = -I_{\sin}$$

since $I_{\sin} = -I_{\sin}$ implies that $I_{\sin}=0$.

Ewald's Sphere

Each X-ray diffraction image represents only a slice, a spherical slice of reciprocal space, as may be seen by the Ewald sphere construction. Both k_{out} and k_{in} have the same length, due to the elastic scattering, since the wavelength has not changed. Therefore, they may be represented as two radial vectors in a sphere in reciprocal space, which shows the values of q that are sampled in a given diffraction image. Since there is a slight spread in the incoming wavelengths of the incoming X-ray beam, the values of $|F(\mathrm{q})|$ can be measured only for q vectors located between the two spheres corresponding to those radii. Therefore, to obtain a full set of Fourier transform data, it is necessary to rotate the crystal through slightly more than 180°, or sometimes less if sufficient symmetry is present. A full 360° rotation is not needed because of a symmetry intrinsic to the Fourier transforms of real functions (such as the electron density), but "slightly more" than 180° is needed to cover all of reciprocal space within a given resolution because of the curvature of the Ewald sphere. In practice, the crystal is rocked by a small amount (0.25-1°) to incorporate reflections near the boundaries of the spherical Ewald shells.

Patterson Function

A well-known result of Fourier transforms is the autocorrelation theorem, which states that the autocorrelation $c(\mathrm{r})$ of a function $f(\mathrm{r})$

$$c(\mathrm{r}) = \int d\mathrm{x} f(\mathrm{x}) f(\mathrm{x}+\mathrm{r}) = \int \frac{d\mathrm{q}}{(2\pi)^3} C(\mathrm{q}) e^{i\mathrm{q}\cdot\mathrm{r}}$$

has a Fourier transform $C(q)$ that is the squared magnitude of $F(q)$

$$C(q) = |F(q)|^2$$

Therefore, the autocorrelation function $c(r)$ of the electron density (also known as the *Patterson function*) can be computed directly from the reflection intensities, without computing the phases. In principle, this could be used to determine the crystal structure directly; however, it is difficult to realise in practice. The autocorrelation function corresponds to the distribution of vectors between atoms in the crystal; thus, a crystal of N atoms in its unit cell may have $N(N-1)$ peaks in its Patterson function. Given the inevitable errors in measuring the intensities, and the mathematical difficulties of reconstructing atomic positions from the interatomic vectors, this technique is rarely used to solve structures, except for the simplest crystals.

Advantages of a Crystal

In principle, an atomic structure could be determined from applying X-ray scattering to non-crystalline samples, even to a single molecule. However, crystals offer a much stronger signal due to their periodicity. A crystalline sample is by definition periodic; a crystal is composed of many unit cells repeated indefinitely in three independent directions. Such periodic systems have a Fourier transform that is concentrated at periodically repeating points in reciprocal space known as *Bragg peaks*; the Bragg peaks correspond to the reflection spots observed in the diffraction image.

Since the amplitude at these reflections grows linearly with the number N of scatterers, the observed *intensity* of these spots should grow quadratically, like N^2. In other words, using a crystal concentrates the weak scattering of the individual unit cells into a much more powerful, coherent reflection that can be observed above the noise. This is an example of constructive interference.

In a liquid, powder or amorphous sample, molecules within that sample are in random orientations. Such samples have a continuous Fourier spectrum that uniformly spreads its amplitude thereby reducing the measured signal intensity, as is observed in SAXS. More importantly, the orientational information is lost.

Although theoretically possible, it is experimentally difficult to obtain atomic-resolution structures of complicated, asymmetric molecules from such rotationally averaged data. An intermediate case is fibre diffraction in which the subunits are arranged periodically in at least one dimension.

X-Ray Scattering Techniques

X-ray scattering techniques are a family of non-destructive analytical techniques which reveal information about the crystal structure, chemical composition, and physical properties of materials and thin films. These techniques are based on observing the scattered intensity of an X-ray beam hitting a sample as a function of incident and scattered angle, polarization, and wavelength or energy. Note that X-ray scattering is different from X-ray diffraction, which is widely used for X-ray crystallography.

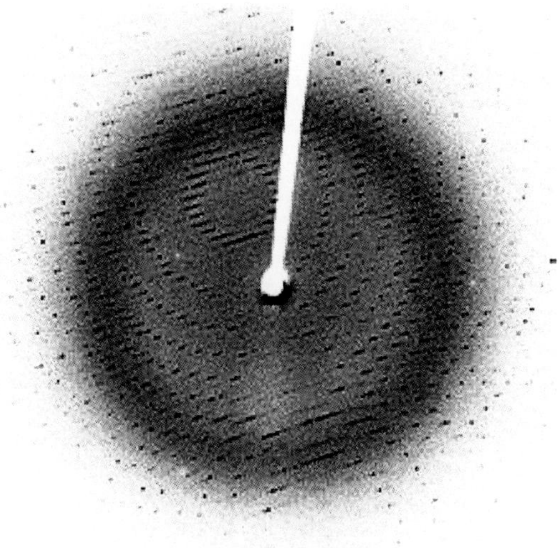

Figure: *This is an X-ray diffraction pattern formed when X-rays are focused on a crystalline material, in this case a protein. Each dot, called a reflection, forms from the coherent interference of scattered X-rays passing through the crystal.*

Scattering Technique

Elastic Scattering: Materials that do not have long range order may also be studied by scattering methods that rely on elastic scattering of monochromatic X-rays.

- Small-angle X-ray scattering (SAXS) probes structure in the nanometre to micrometre range by measuring scattering intensity at scattering angles 2θ close to 0°.
- X-ray reflectivity is an analytical technique for determining thickness, roughness, and density of single layer and multilayer thin films.
- Wide-angle X-ray scattering (WAXS), a technique concentrating on scattering angles 2θ larger than 5°.

Inelastic Scattering: When the energy and angle of the inelastically scattered X-rays are monitored, scattering techniques can be used to probe the electronic band structure of materials. Inelastic scattering alters the phase of the diffracted x-rays, and as a result do not produce useful data for x-ray diffraction. Rather, inelastically scattered x-rays contribute to the background noise in a diffraction pattern.

Micro an Intermediate Step to Nano Level Analysis in Concrete like Composites

Engineered Cementitious Composite (ECC), also called bendable concrete, is an easily molded mortar-based composite reinforced with specially selected short random fibres, usually polymer fibres. Unlike regular concrete, ECC has a strain capacity in the range of 3–7%, compared to 0.1% for ordinary portland cement (OPC). ECC therefore acts more like a ductile metal than a brittle glass (as does OPC concrete), leading to a wide variety of applications.

Development

ECC, unlike common fibre reinforced concrete, is a family of micromechanically designed material. As long as a cementitious material is designed/developed based on micromechanics and fracture mechanics theory to feature large tensile ductility, it can be called an ECC. Therefore, ECC is not a fixed material design, but a broad range of topics under different stages of research, development, and implementations. The ECC material family is expanding. The development of an individual mix design of ECC requires special efforts by systematically engineering of the material at nano-, micro-, macro- and composite scales.

ECC looks similar to ordinary portland cement-based concrete, except that it can deform (or bend) under strain. A number of research groups are developing ECC science, including those at the University of Michigan, University of Houston, Delft University of Technology, the University of Tokyo, the Czech Technical University, and Stanford University. Traditional concrete's lack of durability and failure under strain, both stemming from brittle behavior, have been a pushing factor in the development of ECC.

Properties

ECC has a variety of unique properties, including tensile properties superior to other fibre-reinforced composites, ease of processing on par with conventional cement, the use of only a small volume fraction of fibres (~ 2%), tight crack width, and a lack of anisotropically weak planes. These properties are due largely to the interaction between the fibres and cementing matrix, which can be custom-tailored through micromechanics design. Essentially, the fibres create many microcracks with a very specific width, rather than a few very large cracks (as in conventional concrete.) This allows ECC to deform without catastrophic failure.

This microcracking behavior leads to superior corrosion resistance (the cracks are so small and numerous that it is difficult for aggressive media to penetrate and attack the reinforcing steel) as well as to self-healing. In the presence of water (during a rainstorm, for instance) unreacted cement particles recently exposed due to cracking hydrate and form a number of products (Calcium Silicate Hydrate, calcite, etc.) that expand and fill in the crack. These products appear as a white 'scar' material filling in the crack. This self-healing behavior not only seals the crack to prevent transport of fluids, but mechanical properties are regained. This self-healing has been observed in a variety of conventional cement and concretes; however, above a certain crack width self healing becomes less effective. It is the tightly controlled crack widths seen in ECC that ensure all cracks thoroughly heal when exposed to the natural environment.

When combined with a more conductive material, all cement materials can increase and be used for damage-sensing. This is essentially based on the fact that conductivity will change as damage occurs; the addition of conductive material is meant to raise the conductivity to a level where such changes will be easily identified. Though not a material property of ECC itself, semi-conductive ECC for damage-sensing are being developed.

Types

There are a number of different varieties of ECC, including:

- Lightweight (i.e. low density) ECC have been developed through the addition of air voids, glass bubbles, polymer spheres, and/or lightweight aggregate. Compared to other lightweight concretes, lightweight ECC has superior ductility. Applications include floating homes, barges, and canoes.
- 'Self compacting concrete' refers to a concrete that can flow under its own weight. For instance, a self-compacting material would be able to fill a mold containing elaborate pre-positioned steel reinforcement without the need of vibration or shaking to ensure even distribution. Self-compacting ECC was developed through the use of chemical admixtures to decrease viscosity

and through controlling particle interactions with mix proportioning.

- Sprayable ECC, which can be pneumatically sprayed from a hose, have been developed by using various superplasticizing agents and viscosity-reducing admixtures. Compared to other sprayable fibre-reinforced composites, sprayable ECC has enhanced pumpability in addition to its unique mechanical properties. Sprayable ECC has been used for retrofitting/repair work and tunnel/sewer linings.

An extrudable ECC for use in the extrusion of pipes was first developed in 1998. Extruded ECC pipes have both higher load capacity and higher deformability than any other extruded fibre-reinforced composite pipes.

Field Applications

ECC have found use in a number of large-scale applications in Japan, Korea, Switzerland, Australia and the U.S. These include:

- The Mitaka Dam near Hiroshima was repaired using ECC in 2003. The surface of the then 60-year old dam was severely damaged, showing evidence of cracks, spalling, and some water leakage. A 20 mm-thick layer of ECC was applied by spraying over the 600 m^2 surface.
- Also in 2003, an earth retaining wall in Gifu, Japan, was repaired using ECC. Ordinary portland cement could not be used due to the severity of the cracking in the original structure, which would have caused reflective cracking. ECC was intended to minimize this danger; after one year only microcracks of tolerable width were observed.
- The 95 m (312 ft.) Glorio Roppongi high-rise apartment building in Tokyo contains a total of 54 ECC coupling beams (2 per story) intended to mitigate earthquake damage. The properties of ECC (high damage tolerance, high energy absorption, and ability to deform under shear) give it superior properties in seismic resistance applications when compared to ordinary portland cement. Similar structures include the 41-story Nabeaure Yokohama Tower (4 coupling beams per floor.)
- The 1 km (0.62 mi) long Mihara Bridge in Hokkaido, Japan was opened to traffic in 2005. The steel-reinforced road bed contains nearly 800 m^3 of ECC material. The tensile ductility and tight crack control behavior of ECC led to a 40% reduction in material used during construction.

- Similarly, a 225-mm thick ECC bridge deck on interstate 94 in Michigan was completed in 2005. 30 m^3 of material was used, delivered on-site in standard mixing trucks. Due to the unique mechanical properties of ECC, this deck also used less material than a proposed deck made of ordinary portland cement. Both the University of Michigan and the Michigan Department of Transportation are monitoring the bridge in an attempt to verify the theoretical superior durability of ECC; after 4 years of monitoring, performance remained undiminished.
- The first self-consolidating and high-early-strength ECC patch repair was placed on Ellsworth Road Bridge over US-23 in November 2006. The high-early-strength ECC can achieve a compressive strength of 23.59 ± 1.40 MPa (3422.16 ± 203.33 psi) in 4 hours and 55.59 ± 2.17 MPa (8062.90 ± 315.03 psi) in 28 days, allowing for fast repair and re-opening the session to traffic. The high-early-strength ECC repair has shown superior long-term durability in field conditions compared to typical concrete repair materials.

Comparison to Other Composite Materials

Properties	*FRC*	*Common HPFRCC*	*ECC*
Design Methodology	*N.A.*	***Use high Vf***	***Micromechanics based, minimize Vf for cost and processibility***
Fiber	Any type, Vf usually less than 2%; df for steel ~ 500 micrometre	Mostly steel, Vf usually > 5%; df ~ 150 micrometre	Tailored, polymer fibers, Vf usually less than 2%; df < 50 micrometre
Matrix	Coarse aggregates	Fine aggregates	Controlled for matrix toughness, flaw size; fine sand
Interface	Not controlled	Not controlled	Chemical and frictional bonds controlled for bridging properties
Mechanical Properties	Strain-softening:	Strain-hardening:	Strain-hardening:
Tensile strain	0.1%	<1.5%	>3% (typical); 8% max
Crack width	Unlimited	Typically several hundred micrometres, unlimited beyond 1.5% strain	Typically < 100 micrometres during strain-hardening[1]

Note: FRC=Fibre-Reinforced Cement. HPFRCC=High-Performance Fibre Reinforced Cementitious Composites

High-Performance Fibre-Reinforced Cementitious Composites

High-performance fibre-reinforced cementitious composites (HPFRCCs) are a group of fibre-reinforced cement-based composites which possess the unique ability to flex and self-strengthen before fracturing. This particular class of concrete was developed with the goal of solving the structural problems inherent with today's typical concrete, such as its tendency to fail in a brittle manner under excessive loading and its lack of long-term durability. Because of their design and composition, HPFRCCs possess the remarkable ability to strain harden under excessive loading. In layman's terms, this means they have the ability to flex or deform before fracturing, a behavior similar to that exhibited by most metals under tensile or bending stresses. Because of this capability, HPFRCCs are more resistant to cracking and last considerably longer than normal concrete. Another extremely desirable property of HPFRCCs is their low density. A less dense, and hence lighter material means that HPFRCCs could eventually require much less energy to produce and handle, deeming them a more economic building material. Because of HPFRCCs' lightweight composition and ability to strain harden, it has been proposed that they could eventually become a more durable and efficient alternative to typical concrete.

HPFRCCs are simply a subcategory of ductile fibre-reinforced cementititous composites (DFRCCs) that possess the ability to strain harden under both bending and tensile loads, not to be confused with other DFRCCs that only strain harden under bending loads.

Composition

Because several specific formulas are included in the HPFRCC class, their physical compositions vary considerably. However, most HPFRCCs include at least the following ingredients: fine aggregates, a superplasticizer, polymeric or metallic fibres, cement, and water. Thus the principal difference between HPFRCC and typical concrete composition lies in HPFRCCs' lack of coarse aggregates. Typically, a fine aggregate such as silica sand is used in HPFRCCs.

Material Properties

Strain hardening, the most coveted capability of HPFRCCs, occurs when a material is loaded past its elastic limit and begins to deform plastically. This stretching or 'straining' action actually strengthens the material. This phenomenon is made possible through the development of multiple microscopic cracks, opposed to the single crack/strain softening behavior exhibited by typical fibre-reinforced concretes. It occurs in HPFRCCs as several fibres slip past one another.

One aspect of HPFRCC design involves preventing crack propagation, or the tendency of a crack to increase in length, ultimately leading to material fracture. This occurrence is hindered by the presence of fibre bridging, a property that most HPFRCCs are specifically designed to possess. Fibre bridging is the act of several fibres exerting a force across the width of a crack in an attempt to prevent the crack from developing further. This capability is what gives bendable concrete its ductile properties.

Listed below are some basic mechanical properties of ECC, or Engineered Cementitious Composite, a specific formula of HPFRCC, developed at the University of Michigan. This information is available in Victor C. Li's article on (ECC)- Tailored Composites through Micromechanical Modelling.

The first property listed, the ultimate tensile strength of 4.6 MPa, is slightly larger than the accepted tensile strength of standard fibre-reinforced concretes, (4.3 MPa). More notable, however, is the extremely high ultimate strain value of 5.6% when compared to most FRC's ultimate strain values ranging in the few hundredths of a percent. The first crack stress and first crack strain values are significantly low compared to normal concrete, both the result of the multiple crack phenomenon associated with HPFRCCs.

ECC Material Properties	
Ultimate Tensile Strength (sCU)	4.6 MPa
Ultimate Strain (eCU)	5.6 %
First Crack Stress (sfc)	2.5 MPa
First Crack Strain (efc)	.021 %
Modulus of Elasticity (E)	22 GPa

Design Methodology

The basis for the engineered design of different HPFRCCs varies considerably despite their similar compositions. For instance, the design of one type of HPFRCC called ECC stems from the principles of micromechanics. This field of study is best described as relating macroscopic mechanical properties to a composite's microstructure, and is only one specific method used to design HPFRCCs. Another design methodology used in other formulas of HPFRCCs is based on the material's ability to withstand seismic loading.

Applications

Proposed uses for HPFRCCs include bridge decks, concrete pipes, roads, structures subjected to seismic and non-seismic loads, and other applications where a lightweight, strong and durable building material is desired.

ECC has already been used by the Michigan Department of Transportation to patch a portion of the Grove Street Bridge deck over Interstate 94. The ECC patch was used as a replacement to the previously existent expansion joint that linked two deck slabs. Expansion joints, commonly used in bridges to allow for the seasonal expansion and contraction of the concrete decks, are an example of a ubiquitous construction practice that could eventually be eliminated through the use of bendable concrete.

Other existent structures composed of HPFRCCs, specifically ECC, include the Curtis Road Bridge in Ann Arbor, MI and the Mihara Bridge in Hokkaido, Japan. The deck of the Mihara Bridge, composed of bendable concrete, is only five centimetres thick and has an expected lifetime of one-hundred years.

Though HPFRCCs have been tested extensively in the lab and been employed in a few commercial building projects, further long-term research and real-world application is needed to prove the true benefits of this material.

6

Moving Window Representation of Interfacial Debonding in Concrete

Concrete is a composite material composed of water, coarse granular material (the fine and coarse aggregate or filler) embedded in a hard matrix of material (the cement or binder) that fills the space among the aggregate particles and glues them together.

Famous concrete structures include the Hoover Dam, the Panama Canal and the Roman Pantheon. The earliest large-scale users of concrete technology were the ancient Romans, and concrete was widely used in the Roman Empire. The Colosseum in Rome was built largely of concrete, and the concrete dome of the Pantheon is the world's largest unreinforced concrete dome. After the Roman Empire collapsed, use of concrete became rare until the technology was re-pioneered in the mid-18th century. Today, concrete is the most widely used man-made material (measured by tonnage).

Composition of Concrete

There are many types of concrete available, created by varying the proportions of the main ingredients below. In this way or by substitution for the cementitious and aggregate phases, the finished product can be tailored to its application with varying strength, density, or chemical and thermal resistance properties.

"Aggregate" consists of large chunks of material in a concrete mix, generally a coarse gravel or crushed rocks such as limestone, or granite, along with finer materials such as sand. "Cement", commonly Portland cement, and other cementitious materials such as fly ash and slag cement, serve as a binder for the aggregate.

Water is then mixed with this dry composite, which produces a semi-liquid that workers can shape (typically by pouring it into a form). The concrete solidifies and hardens to rock-hard strength through a chemical process called hydration. The water reacts with the cement, which bonds the other components together, creating a robust stone-like material.

"Chemical admixtures" are added to achieve varied properties. These ingredients may speed or slow down the rate at which the concrete hardens, and impart many other useful properties.

"Reinforcements" are often added to concrete. Concrete can be formulated with high compressive strength, but always has lower tensile strength. For this reason it is usually reinforced with materials that are strong in tension (often steel).

"Mineral admixtures" are becoming more popular in recent decades. The use of recycled materials as concrete ingredients has been gaining popularity because of increasingly stringent environmental legislation, and the discovery that such materials often have complementary and valuable properties.

The most conspicuous of these are fly ash, a by-product of coal-fired power plants, and silica fume, a byproduct of industrial electric arc furnaces. The use of these materials in concrete reduces the amount of resources required, as the ash and fume act as a cement replacement. This displaces some cement production, an energetically expensive and environmentally problematic process, while reducing the amount of industrial waste that must be disposed of.

The *mix design* depends on the type of structure being built, how the concrete is mixed and delivered, and how it is placed to form the structure.

Cement

Portland cement is the most common type of cement in general usage. It is a basic ingredient of concrete, mortar and plaster. English masonry worker Joseph Aspdin patented Portland cement in 1824.

It was named because of the similarity of its colour to Portland limestone, quarried from the English Isle of Portland and used extensively in London architecture. It consists of a mixture of oxides of calcium, silicon and aluminium. Portland cement and similar materials are made by heating limestone (a source of calcium) with clay and grinding this product (called *clinker*) with a source of sulfate (most commonly gypsum).

***Figure:** A few tons of bagged cement. This amount represents about two minutes of output from a 10,000 ton per day cement kiln.*

In modern cement kilns many advanced features are used to lower the fuel consumption per ton of clinker produced. Cement kilns are extremely large, complex, and inherently dusty industrial installations, and have emissions which must be controlled.

Of the various ingredients used in concrete the cement is the most energetically expensive. Even complex and efficient kilns require 3.3 to 3.6 gigajoules of energy to produce a ton of clinker and then grind it into cement. Many kilns can be fuelled with difficult-to-dispose-of wastes, the most common being used tires.

The extremely high temperatures and long periods of time at those temperatures allows cement kilns to efficiently and completely burn even difficult-to-use fuels.

Water

Combining water with a cementitious material forms a cement paste by the process of hydration.

The cement paste glues the aggregate together, fills voids within it, and makes it flow more freely. A lower water-to-cement ratio yields a stronger, more durable concrete, whereas more water gives a freer-flowing concrete with a higher slump. Impure water used to make concrete can cause problems when setting or in causing premature failure of the structure.

Hydration involves many different reactions, often occurring at the same time. As the reactions proceed, the products of the cement hydration process gradually bond together the individual sand and gravel particles and other components of the concrete to form a solid mass.

Reaction:

Cement chemist notation: $C_3S + H \rightarrow C\text{-}S\text{-}H + CH$

Standard notation: $Ca_3SiO_5 + H_2O \rightarrow (CaO)\cdot(SiO_2)\cdot(H_2O)(gel) + Ca(OH)_2$

Balanced: $2Ca_3SiO_5 + 7H_2O \rightarrow 3(CaO)\cdot 2(SiO_2)\cdot 4(H_2O)(gel) + 3Ca(OH)_2$

Aggregates

Figure: *Crushed stone aggregate*

Fine and coarse aggregates make up the bulk of a concrete mixture. Sand, natural gravel and crushed stone are used mainly for this purpose. Recycled aggregates (from construction, demolition and excavation waste) are increasingly used as partial replacements of natural aggregates, while a number of manufactured aggregates, including air-cooled blast furnace slag and bottom ash are also permitted.

The presence of aggregate greatly increases the durability of concrete above that of cement, which is a brittle material in its pure state. Thus concrete is a true composite material.

Redistribution of aggregates after compaction often creates inhomogeneity due to the influence of vibration. This can lead to strength gradients.

Decorative stones such as quartzite, small river stones or crushed glass are sometimes added to the surface of concrete for a decorative "exposed aggregate" finish, popular among landscape designers.

In addition to being decorative, exposed aggregate adds robustness to a concrete driveway.

Reinforcement

Concrete is strong in compression, as the aggregate efficiently carries the compression load. However, it is weak in tension as the cement holding the aggregate in place can crack, allowing the structure to fail. Reinforced concrete adds either steel reinforcing bars, steel fibres, glass fibre, or plastic fibre to carry tensile loads.

Figure: *Constructing a rebar cage. This cage will be permanently embedded in poured concrete to create a reinforced concrete structure.*

Chemical Admixtures

Chemical admixtures are materials in the form of powder or fluids that are added to the concrete to give it certain characteristics not obtainable with plain concrete mixes. In normal use, admixture dosages are less than 5% by mass of cement and are added to the concrete at the time of batching/mixing. The common types of admixtures are as follows.

- Accelerators speed up the hydration (hardening) of the concrete. Typical materials used are $CaCl_2$, $Ca(NO_3)_2$ and $NaNO_3$. However, use of chlorides may cause corrosion in steel reinforcing and is prohibited in some countries, so that nitrates may be favoured.
- Retarders slow the hydration of concrete and are used in large or difficult pours where partial setting before the pour is complete is undesirable. Typical polyol retarders are sugar, sucrose, sodium gluconate, glucose, citric acid, and tartaric acid.
- Air entrainments add and entrain tiny air bubbles in the concrete, which reduces damage during freeze-thaw cycles, increasing durability. However, entrained air entails a trade off with strength, as each 1% of air may decrease compressive strength 5%.

- Plasticizers increase the workability of plastic or "fresh" concrete, allowing it be placed more easily, with less consolidating effort. A typical plasticizer is lignosulfonate. Plasticizers can be used to reduce the water content of a concrete while maintaining workability and are sometimes called *water-reducers* due to this use. Such treatment improves its strength and durability characteristics. Superplasticizers (also called *high-range water-reducers*) are a class of plasticizers that have fewer deleterious effects and can be used to increase workability more than is practical with traditional plasticizers. Compounds used as superplasticizers include sulfonated naphthalene formaldehyde condensate, sulfonated melamine formaldehyde condensate, acetone formaldehyde condensate and polycarboxylate ethers.
- Pigments can be used to change the colour of concrete, for aesthetics.
- Corrosion inhibitors are used to minimize the corrosion of steel and steel bars in concrete.
- Bonding agents are used to create a bond between old and new concrete (typically a type of polymer) .
- Pumping aids improve pumpability, thicken the paste and reduce separation and bleeding.

Mineral Admixtures and Blended Cements

There are inorganic materials that also have pozzolanic or latent hydraulic properties. These very fine-grained materials are added to the concrete mix to improve the properties of concrete (mineral admixtures), or as a replacement for Portland cement (blended cements). Products which incorporate limestone, fly ash, blast furnace slag, and other useful materials with pozzolanic properties into the mix, are being tested and used. This development is due to cement production being one of the largest producers (at about 5 to 10%) of global greenhouse gas emissions, as well as lowering costs, improving concrete properties, and recycling wastes.

- Fly ash: A by-product of coal-fired electric generating plants, it is used to partially replace Portland cement (by up to 60% by mass). The properties of fly ash depend on the type of coal burnt. In general, siliceous fly ash is pozzolanic, while calcareous fly ash has latent hydraulic properties.
- Ground granulated blast furnace slag (GGBFS or GGBS): A by-product of steel production is used to partially replace Portland cement (by up to 80% by mass). It has latent hydraulic properties.

- Silica fume: A byproduct of the production of silicon and ferrosilicon alloys. Silica fume is similar to fly ash, but has a particle size 100 times smaller. This results in a higher surface-to-volume ratio and a much faster pozzolanic reaction. Silica fume is used to increase strength and durability of concrete, but generally requires the use of superplasticizers for workability.
- High reactivity Metakaolin (HRM): Metakaolin produces concrete with strength and durability similar to concrete made with silica fume. While silica fume is usually dark gray or black in colour, high-reactivity metakaolin is usually bright white in colour, making it the preferred choice for architectural concrete where appearance is important.

Concrete Production

Figure: *Concrete plant facility showing a Concrete mixer being filled from the ingredient silos.*

Concrete production is the process of mixing together the various ingredients—water, aggregate, cement, and any additives—to produce concrete. Concrete production is time-sensitive. Once the ingredients are mixed, workers must put the concrete in place before it hardens. In modern usage, most concrete production takes place in a large type of industrial facility called a concrete plant, or often a batch plant.

In general usage, concrete plants come in two main types, ready mix plants and central mix plants. A ready mix plant mixes all the ingredients except water, while a central mix plant mixes all the ingredients including water. A central mix plant offers more accurate control of the concrete quality through better measurements of the amount of water added, but must be placed closer to the work site where the concrete will be used, since hydration begins at the plant.

A concrete plant consists of large storage hoppers for various reactive ingredients like cement, storage for bulk ingredients like aggregate and water, mechanisms for the addition of various additives and amendments, machinery to accurately weigh, move, and mix some or all of those ingredients, and facilities to dispense the mixed concrete, often to a concrete mixer truck. Modern concrete is usually prepared as a viscous fluid, so that it may be poured into forms, which are containers erected in the field to give the concrete its desired shape. There are many different ways in which concrete formwork can be prepared, such as Slip forming and Steel plate construction. Alternatively, concrete can be mixed into dryer, non-fluid forms and used in factory settings to manufacture Precast concrete products.

There is a wide variety of equipment for processing concrete, from hand tools to heavy industrial machinery. Whichever equipment builders use, however, the objective is to produce the desired building material; ingredients must be properly mixed, placed, shaped, and retained within time constraints. Once the mix is where it should be, the curing process must be controlled to ensure that the concrete attains the desired attributes. During concrete preparation, various technical details may affect the quality and nature of the product.

When initially mixed, Portland cement and water rapidly form a gel of tangled chains of interlocking crystals, and components of the gel continue to react over time. Initially the gel is fluid, which improves workability and aids in placement of the material, but as the concrete sets, the chains of crystals join into a rigid structure, counteracting the fluidity of the gel and fixing the particles of aggregate in place. During curing, the cement continues to react with the residual water in a process of hydration. In properly formulated concrete, once this curing process has terminated the product has the desired physical and chemical properties. Among the qualities typically desired, are mechanical strength, low moisture permeability, and chemical and volumetric stability.

Mixing Concrete

Thorough mixing is essential for the production of uniform, high-quality concrete. For this reason equipment and methods should be capable of effectively mixing concrete materials containing the largest specified aggregate to produce *uniform mixtures* of the lowest slump practical for the work.

Separate paste mixing has shown that the mixing of cement and water into a paste before combining these materials with aggregates can increase the compressive strength of the resulting concrete. The paste is generally mixed in a *high-speed*, shear-type mixer at a w/cm (water to cement ratio) of 0.30 to 0.45 by mass. The cement paste premix may include admixtures such as accelerators or retarders, superplasticizers, pigments, or silica fume. The premixed paste is then blended with aggregates and any remaining batch water and final mixing is completed in conventional concrete mixing equipment.

High-energy mixed (HEM) concrete is produced by means of high-speed mixing of cement, water and sand with net specific energy consumption of at least 5 kilojoules per kilogram of the mix. A plasticizer or a superplasticizer is then added to the activated mixture, which can later be mixed with aggregates in a conventional concrete mixer. In this process, sand provides dissipation of energy and creates high-shear conditions on the surface of cement particles. This results in the full volume of water interacting with cement. The liquid activated mixture can be used by itself or foamed (expanded) for lightweight concrete. HEM concrete hardens in low and subzero temperature conditions and possesses an increased volume of gel, which drastically reduces capillarity in solid and porous materials.

Workability

Figure: *Pouring and smoothing out concrete at Palisades Park in Washington DC.*

Workability is the ability of a fresh (plastic) concrete mix to fill the form/mold properly with the desired work (vibration) and without reducing the concrete's quality. Workability depends on water content, aggregate (shape and size distribution), cementitious content and age (level of hydration) and can be modified by adding chemical admixtures, like superplasticizer. Raising the water content or adding chemical admixtures increases concrete workability. Excessive water leads to increased bleeding (surface water) and/or segregation of aggregates (when the cement and aggregates start to separate), with the resulting concrete having reduced quality. The use of an aggregate with an undesirable gradation can result in a very harsh mix design with a very low slump, which cannot readily be made more workable by addition of reasonable amounts of water.

Workability can be measured by the concrete slump test, a simplistic measure of the plasticity of a fresh batch of concrete following the ASTM C 143 or EN 12350-2 test standards. Slump is normally measured by filling an "Abrams cone" with a sample from a fresh batch of concrete. The cone is placed with the wide end down onto a level, non-absorptive surface. It is then filled in three layers of equal volume, with each layer being tamped with a steel rod to consolidate the layer. When the cone is carefully lifted off, the enclosed material slumps a certain amount, owing to gravity. A relatively dry sample slumps very little, having a slump value of one or two inches (25 or 50 mm) out of one foot (305 mm). A relatively wet concrete sample may slump as much as eight inches. Workability can also be measured by the flow table test.

Slump can be increased by addition of chemical admixtures such as plasticizer or superplasticizer without changing the water-cement ratio. Some other admixtures, especially air-entraining admixture, can increase the slump of a mix.

High-flow concrete, like self-consolidating concrete, is tested by other flow-measuring methods. One of these methods includes placing the cone on the narrow end and observing how the mix flows through the cone while it is gradually lifted.

After mixing, concrete is a fluid and can be pumped to the location where needed.

Curing

In all but the least critical applications, care must be taken to properly *cure* concrete, to achieve best strength and hardness. This happens after the concrete has been placed. Cement requires a moist, controlled environment to gain strength and harden fully. The cement paste hardens over time, initially setting and becoming rigid though very weak and gaining in strength in the weeks following. In around 4 weeks, typically over 90% of the final strength is reached, though strengthening may continue for decades. The conversion of calcium hydroxide in the concrete into calcium carbonate from absorption of CO_2 over several decades further strengthens the concrete and makes it more resistant to damage. However, this reaction, called carbonation, lowers the pH of the cement pore solution and can cause the reinforcement bars to corrode.

Figure: *A concrete slab ponded while curing.*

Hydration and hardening of concrete during the first three days is critical. Abnormally fast drying and shrinkage due to factors such as evaporation from wind during placement may lead to increased tensile stresses at a time when it has not yet gained sufficient strength, resulting in greater shrinkage cracking. The early strength of the concrete can be increased if it is kept damp during the curing process. Minimizing stress prior to curing minimizes cracking. High-early-strength concrete is designed to hydrate faster, often by increased use of cement that increases shrinkage and cracking. The strength of concrete changes (increases) for up to three years. It depends on cross-section dimension of elements and conditions of structure exploitation.

During this period concrete must be kept under controlled temperature and humid atmosphere. In practice, this is achieved by spraying or ponding the concrete surface with water, thereby protecting the concrete mass from ill effects of ambient conditions. The picture to the right shows one of many ways to achieve this, ponding – submerging setting concrete in water and wrapping in plastic to contain the water in the mix. Additional common curing methods include wet burlap and/or plastic sheeting covering the fresh concrete, or by spraying on a water-impermeable temporary curing membrane.

Properly curing concrete leads to increased strength and lower permeability and avoids cracking where the surface dries out prematurely. Care must also be taken to avoid freezing or overheating due to the exothermic setting of cement. Improper curing can cause scaling, reduced strength, poor abrasion resistance and cracking.

Properties

Concrete has relatively high compressive strength, but much lower tensile strength. For this reason it is usually reinforced with materials that are strong in tension (often steel). The elasticity of concrete is relatively constant at low stress levels but starts decreasing at higher stress levels as matrix cracking develops. Concrete has a very low coefficient of thermal expansion and shrinks as it matures. All concrete structures crack to some extent, due to shrinkage and tension. Concrete that is subjected to long-duration forces is prone to creep.

Tests can be performed to ensure that the properties of concrete correspond to specifications for the application.

Different mixes of concrete ingredients produce different strengths, which are measured in psi or MPa.

Different strengths of concrete are used for different purposes. Very low-strength (2000 psi or less) concrete may be used when the concrete must be lightweight. Lightweight concrete is often achieved by adding air, foams, or lightweight aggregates, with the side effect that the strength is reduced. For most routine uses, 3000-psi to 4000-psi concrete is often used. 5000-psi concrete is readily commercially available as a more durable, although more expensive, option. 5000-psi concrete is often used for larger civil projects. Strengths above 5000 psi are often used for specific building elements. For example, the lower floor columns of high-rise concrete buildings may use concrete of 12,000 psi or more, to keep the size of the columns small. Bridges may use long beams of 10,000 psi concrete to lower the number of spans required. Occasionally, other structural needs may require high-strength concrete. If a structure must be very rigid, concrete of very high strength may be specified, even much stronger than is required to bear the service loads. Strengths as high as 19,000 psi have been used commercially for these reasons.

Figure: *Compression testing of a concrete cylinder*

Imperial Strength	*Metric Equivalent*
2000 psi	14 MPa
2500 psi	18 MPa
3000 psi	20 MPa
3500 psi	25 MPa
4000 psi	30 MPa
5000 psi	35 MPa
6000 psi	40 MPa
7000 psi	50 MPa
8000 psi	55 MPa
10000 psi	70 MPa
12000 psi	80 MPa
19000 psi	130 MPa
36000 psi	250 MPa

Concrete Degradation

Concrete can be damaged by many processes, such as the expansion of corrosion products of the steel reinforcement bars, freezing of trapped water, fire or radiant heat, aggregate expansion, sea water effects, bacterial corrosion, leaching, erosion by fast-flowing water, physical damage and chemical damage (from carbonatation, chlorides, sulfates and distillate water). The micro fungi Aspergillus Alternaria and Cladosporium were able to grow on samples of concrete used as a radioactive waste barrier in the Chernobyl reactor; leaching aluminium, iron, calcium and silicon.

Figure: *Concrete spalling caused by the corrosion of rebar*

Microbial Concrete

Bacteria such as Bacillus pasteurii, Bacillus pseudofirmus, Bacillus cohnii, Sporosarcina pasteuri, and Arthrobacter crystallopoietes increase the compression strength of concrete through their biomass. Not all bacteria increase the strength of concrete significantly with their biomass. Bacillus sp. CT-5.can reduce corrosion of reinforcement in reinforced concrete by up to four times. Sporosarcina pasteuriireduced reduces water and chloride permeability. B. pasteurii increases resistance to acid. Bacillus pasteurii and B. sphaericuscan induce calcium carbonate precipitation in the surface of cracks, adding compression strength.

Environmental and Health

The manufacture and use of concrete produce a wide range of environmental and social consequences. Some are harmful, some welcome, and some both, depending on circumstances. A major component of concrete is cement, which similarly exerts environmental and social effects.

The cement industry is one of the three primary producers of carbon dioxide, a major greenhouse gas. The other two are the energy production and transportation industries. As of 2011 it contributes 7%

to global anthropogenic CO2 emissions; largely due to the sintering of limestone and clay at 1500 C.

Concrete is used to create hard surfaces that contribute to surface runoff, which can cause heavy soil erosion, water pollution, and flooding, but conversely can be used to divert, dam, and control flooding. Concrete is a primary contributor to the urban heat island effect, though less so than asphalt. Workers who cut, grind or polish concrete are at risk of inhaling airborne silica, which can lead to silicosis. Concrete dust released by building demolition and natural disasters can be a major source of dangerous air pollution.

The presence of some substances in concrete, including useful and unwanted additives, can cause health concerns due to toxicity and radioactivity. Wet concrete is highly alkaline and must be handled with proper protective equipment.

Figure: *Recycled crushed concrete, to be reused as granular fill, is loaded into a semi-dump truck.*

Concrete Recycling

Concrete recycling is an increasingly common method of disposing of concrete structures. Concrete debris was once routinely shipped to landfills for disposal, but recycling is increasing due to improved environmental awareness, governmental laws and economic benefits.

Concrete, which must be free of trash, wood, paper and other such materials, is collected from demolition sites and put through a crushing machine, often along with asphalt, bricks and rocks.

Reinforced concrete contains rebar and other metallic reinforcements, which are removed with magnets and recycled elsewhere. The remaining aggregate chunks are sorted by size. Larger chunks may go through the crusher again. Smaller pieces of concrete are used as gravel for new construction projects. Aggregate base gravel is laid down as the lowest layer in a road, with fresh concrete or asphalt placed over it. Crushed recycled concrete can sometimes be used as the dry aggregate for brand new concrete if it is free of contaminants, though the use of recycled concrete limits strength and is not allowed in many jurisdictions. On 3 March 1983, a government-funded research team (the VIRL research.codep) estimated that almost 17% of worldwide landfill was by-products of concrete based waste.

Use of Concrete in Infrastructure

Figure: *Aerial photo of reconstruction at Taum Sauk (Missouri) pumped storage facility in late November, 2009. After the original reservoir failed, the new reservoir was made of roller-compacted concrete.*

Mass Concrete Structures

Large concrete structures such as dams, navigation locks, large mat foundations, and large breakwaters generate excessive heat during cement hydration and associated expansion. To mitigate these effects post-cooling is commonly applied during construction. An early example at Hoover Dam, installed a network of pipes between vertical concrete placements to circulate cooling water during the curing process to avoid damaging overheating. Similar systems are still used; depending on volume of the pour, the concrete mix used, and ambient air temperature, the cooling process may last for many months after the concrete is placed. Various methods also are used to pre-cool the concrete mix in mass concrete structures.

Another approach to mass concrete structures that is becoming more widespread is the use of roller-compacted concrete, which uses much lower amounts of cement and water than conventional concrete mixtures and is generally not poured into place. Instead it is placed in thick layers as a semi-dry material and compacted into a dense, strong mass with rolling compactors. Because it uses less cementitious material, roller-compacted concrete has a much lower cooling requirement than conventional concrete.

Prestressed Concrete Structures

Prestressed concrete is a form of reinforced concrete that builds in compressive stresses during construction to oppose those experienced in use. This can greatly reduce the weight of beams or slabs, by better distributing the stresses in the structure to make optimal use of the reinforcement. For example, a horizontal beam tends to sag. Prestressed reinforcement along the bottom of the beam counteracts this. In pre-tensioned concrete, the prestressing is achieved by using steel or polymer tendons or bars that are subjected to a tensile force prior to casting, or for post-tensioned concrete, after casting.

Concrete Textures

When one thinks of concrete, the image of a dull, gray concrete wall often comes to mind. With the use of form liner, concrete can be cast and molded into different textures and used for decorative concrete applications. Sound/retaining walls, bridges, office buildings and more serve as the optimal canvases for concrete art. For example, the Pima Freeway/Loop 101 retaining and sound walls in Scottsdale, Arizona, feature desert flora and fauna, a 67-foot (20 m) lizard and 40-foot (12 m) cacti along the 8-mile (13 km) stretch. The project, titled "The Path Most Travelled," is one example of how concrete can be shaped using elastomeric form liner.

Building with Concrete

Concrete is one of the most durable building materials. It provides superior fire resistance compared with wooden construction and gains strength over time. Structures made of concrete can have a long service life. Concrete is used more than any other manmade material in the world. As of 2006, about 7.5 billion cubic metres of concrete are made each year, more than one cubic metre for every person on Earth.

More than 55,000 miles (89,000 km) of highways in the United States are paved with this material. Reinforced concrete, prestressed concrete and precast concrete are the most widely used types of concrete functional extensions in modern days.

Energy Efficiency

Energy requirements for transportation of concrete are low because it is produced locally from local resources, typically manufactured within 100 kilometres of the job site. Similarly, relatively little energy is used in producing and combining the raw materials (although large amounts of CO_2 are produced by the chemical reactions in cement manufacture). The overall embodied energy of concrete is therefore lower than for most structural materials other than wood.

Once in place, concrete offers great energy efficiency over the lifetime of a building. Concrete walls leak air far less than those made of wood frames. Air leakage accounts for a large percentage of energy loss from a home. The thermal mass properties of concrete increase the efficiency of both residential and commercial buildings. By storing and releasing the energy needed for heating or cooling, concrete's thermal mass delivers year-round benefits by reducing temperature swings inside and minimizing heating and cooling costs. While insulation reduces energy loss through the building envelope, thermal mass uses walls to store and release energy. Modern concrete wall systems use both external insulation and thermal mass to create an energy-efficient building. Insulating concrete forms (ICFs) are hollow blocks or panels made of either insulating foam or rastra that are stacked to form the shape of the walls of a building and then filled with reinforced concrete to create the structure.

Pervious Concrete

Pervious concrete is a mix of specially graded coarse aggregate, cement, water and little-to-no fine aggregates. This concrete is also known as "no-fines" or porous concrete. Mixing the ingredients in a carefully controlled process creates a paste that coats and bonds the aggregate particles. The hardened concrete contains interconnected air voids totalling approximately 15 to 25 percent. Water runs through the voids in the pavement to the soil underneath. Air entrainment admixtures are often used in freeze–thaw climates to minimize the possibility of frost damage.

Nano Concrete

Concrete is the most widely manufactured construction material. The addition of carbon nanofibres to concrete has many advantages in terms of mechanical and electrical properties (e.g. higher strength and higher Young's modulus) and self-monitoring behavior due to the high tensile strength and high conductivity. Mullapudi used the pulse velocity method to characterize the properties of concrete containing carbon nanofibres. The test results indicate that the compressive strength and percentage reduction in electrical resistance while loading concrete containing carbon nanofibres differ from those of plain concrete. A reasonable concentration of carbon nanofibres need to be determined for use in concrete, which not only enhances compressive strength, but also improves the electrical properties required for strain monitoring, damage evaluation and self-health monitoring of concrete.

Fire Safety

Concrete buildings are more resistant to fire than those constructed using steel frames, since concrete has lower heat conductivity than steel and can thus last longer under the same fire conditions. Concrete is sometimes used as a fire protection for steel frames, for the same effect as above. Concrete as a fire shield, for example Fondu fyre, can also be used in extreme environments like a missile launch pad.

Options for non-combustible construction include floors, ceilings and roofs made of cast-in-place and hollow-core precast concrete. For walls, concrete masonry technology and Insulating Concrete Forms (ICFs) are additional options. ICFs are hollow blocks or panels made of fireproof insulating foam that are stacked to form the shape of the walls of a building and then filled with reinforced concrete to create the structure.

Concrete also provides good resistance against externally applied forces such as high winds, hurricanes, and tornadoes owing to its lateral stiffness, which results in minimal horizontal movement. However this stiffness can work against certain types of concrete structures, particularly where a relatively higher flexing structure is require to resist more extreme forces.

Earthquake Safety

As discussed above, concrete is very strong in compression, but weak in tension. Larger earthquakes can generate very large shear loads on structures. These shear loads subject the structure to both tensile and compressional loads. Concrete structures without reinforcement, like other unreinforced masonry structures, can fail during severe earthquake shaking. Unreinforced masonry structures constitute one of the largest earthquake risks globally. These risks can be reduced through seismic retrofitting of at-risk buildings, (e.g. school buildings in Istanbul, Turkey).

Useful Life

Concrete can be viewed as a form of artificial sedimentary rock. As a type of mineral, the compounds of which it is composed are extremely stable. Many concrete structures are built with an expected lifetime of approximately 100 years, but researchers have suggested that adding silica fume could extend the useful life of bridges and other concrete uses to as long as 16,000 years. Coatings are also available to protect concrete from damage, and extend the useful life. Epoxy coatings may be applied only to interior surfaces, though, as they would otherwise trap moisture in the concrete.

A self-healing concrete has been developed that can also last longer than conventional concrete.

Large dams, such as the Hoover Dam, and the Three Gorges Dam are intended to last "forever", a period that is not quantified.

World Records

The world record for the largest concrete pour in a single project is the Three Gorges Dam in Hubei Province, China by the Three Gorges Corporation. The amount of concrete used in the construction of the dam is estimated at 16 million cubic metres over 17 years. The previous record was 12.3 million cubic metres held by Itaipu hydropower station in Brazil.

The world record for concrete pumping was set on 7 August 2009 during the construction of the Parbati Hydroelectric Project, near the village of Suind, Himachal Pradesh, India, when the concrete mix was pumped through a vertical height of 715 m (2,346 ft).

The world record for the largest continuously poured concrete raft was achieved in August 2007 in Abu Dhabi by contracting firm Al Habtoor-CCC Joint Venture and the concrete supplier is Unibeton Ready Mix. The pour (a part of the foundation for the Abu Dhabi's Landmark Tower) was 16,000 cubic metres of concrete poured within a two-day period. The previous record, 13,200 cubic metres poured in 54 hours despite a severe tropical storm requiring the site to be covered with tarpaulins to allow work to continue, was achieved in 1992 by joint Japanese and South Korean consortiums Hazama Corporation and the Samsung C&T Corporation for the construction of the Petronas Towers in Kuala Lumpur, Malaysia.

The world record for largest continuously poured concrete floor was completed 8 November 1997, in Louisville, Kentucky by design-build firm EXXCEL Project Management. The monolithic placement consisted of 225,000 square feet (20,900 m^2) of concrete placed within a 30-hour period, finished to a flatness tolerance of F_F 54.60 and a levelness tolerance of F_L 43.83. This surpassed the previous record by 50% in total volume and 7.5% in total area.

The record for the largest continuously placed underwater concrete pour was completed 18 October 2010, in New Orleans, Louisiana by contractor C. J. Mahan Construction Company, LLC of Grove City, Ohio. The placement consisted of 10,251 cubic yards of concrete placed in a 58.5 hour period using two concrete pumps and two dedicated concrete batch plants. Upon curing, this placement allows the 50,180-square-foot (4,662 m^2) cofferdam to be dewatered approximately 26 feet (7.9 m) below sea level to allow the construction of the Inner Harbor Navigation Canal Sill & Monolith Project to be completed in the dry.

Molecular Modelling of Confined Fluids and Solidfluid Interfaces in Portland

Portland cement is the most common type of cement in general use around the world, used as a basic ingredient of concrete, mortar, stucco, and most non-speciality grout. It usually originates from limestone. It is a fine powder produced by grinding Portland cement clinker (more than 90%), a limited amount of calcium sulfate (which controls the set time) and up to 5% minor constituents as allowed by various standards such as the European Standard EN 197-1:

Portland cement clinker is a hydraulic material which shall consist of at least two-thirds by mass of calcium silicates (3 $CaO \cdot SiO_2$ and 2 $CaO \cdot SiO_2$), the remainder consisting of aluminium- and iron-containing clinker phases and other compounds. The ratio of CaO to SiO_2 shall not be less than 2.0. The magnesium oxide content (MgO) shall not exceed 5.0% by mass.

ASTM C150 defines Portland cement as "hydraulic cement (cement that not only hardens by reacting with water but also forms a water-resistant product) produced by pulverizing clinkers consisting essentially of hydraulic calcium silicates, usually containing one or more of the forms of calcium sulfate as an inter ground addition."

Clinkers are nodules (diameters, 0.2–1.0 inch [5–25 mm]) of a sintered material that is produced when a raw mixture of predetermined composition is heated to high temperature.

The low cost and widespread availability of the limestone, shales, and other naturally occurring materials make Portland cement one of the lowest-cost materials widely used over the last century throughout the world. Concrete is one of the most versatile construction materials available in the world.

Portland cement clinker is made by heating, in a kiln, a homogeneous mixture of raw materials to a calcining temperature, which is about 1450°C for modern cements. The aluminium oxide and iron oxide are present as a flux and contribute little to the strength. For special cements, such as Low Heat (LH) and Sulfate Resistant (SR) types, it is necessary to limit the amount of tricalcium aluminate (3 CaO $\cdot Al_2O_3$) formed. The major raw material for the clinker-making is usually limestone ($CaCO_3$) mixed with a second material containing clay as source of alumino-silicate. Normally, an impure limestone which contains clay or SiO_2 is used. The $CaCO_3$ content of these limestones can be as low as 80%. Secondary raw materials (materials in the rawmix other than limestone) depend on the purity of the limestone. Some of the materials used are clay, shale, sand, iron ore, bauxite, fly ash, and slag. When a cement kiln is fired by coal, the ash of the coal acts as a secondary raw material.

Portland cement was developed from natural cements made in Britain in the early part of the nineteenth century, and its name is derived from its similarity to Portland stone, a type of building stone that was quarried on the Isle of Portland in Dorset, England. Joseph Aspdin, a British bricklayer from Leeds, is considered to be one of the originators of Portland cement after Louis Vicat in 1817. A process for the manufacture of Portland cement was patented in 1824. This cement was an artificial cement similar in properties to the material known as "Roman cement", which had been patented in 1796 by James Parker. Aspdin's process was similar to a process patented in 1822 and used since 1811 by James Frost who called his cement "British cement". The name "Portland cement" is also recorded in a directory published in 1823 being associated with a William Lockwood, Dave Stewart, and possibly others.

James Frost is reported to have erected a manufactory for making of an artificial cement in 1826. In 1843, Aspdin's son William improved their cement, which was initially called "Patent Portland cement", although he had no patent. In 1848, William Aspdin further improved his cement; in 1853, he moved to Germany where he was involved in cement making. William Aspdin made what could be called meso-Portland cement (a mix of Portland cement and hydraulic lime). John Grant of the Metropolitan Board of Works in 1859 set out requirements for cement to be used in the London sewer project. This became a specification for Portland cement. The Hoffman "endless" kiln which gave "perfect control over combustion" was tested in 1860 and showed the process produced a better grade of cement. This cement was made at the Portland Cementfabrik Stern at Stettin, which was the first to utilise a Hoffman kiln. It is thought that the first modern Portland cement was made there. The Association of German Cement Manufacturers issued a standard on Portland cement in 1878.

Cement Grinding

In order to achieve the desired setting qualities in the finished product, a quantity (2–8%, but typically 5%) of calcium sulfate (usually gypsum or anhydrite) is added to the clinker and the mixture is finely ground to form the finished cement powder. This is achieved in a cement mill. The grinding process is controlled to obtain a powder with a broad particle size range, in which typically 15% by mass consists of particles below 5 μm diameter, and 5% of particles above 45 μm. The measure of fineness usually used is the "specific surface area", which is the total particle surface area of a unit mass of cement. The rate of initial reaction (up to 24 hours) of the cement on addition of water is directly proportional to the specific surface area. Typical values are 320–380 $m^2 \cdot kg^{-1}$ for general purpose cements, and 450–650 $m^2 \cdot kg^{-1}$ for "rapid hardening" cements. The cement is conveyed by belt or powder pump to a silo for storage. Cement plants normally have sufficient silo space for 1–20 weeks production, depending upon local demand cycles. The cement is delivered to end-users either in bags or as bulk powder blown from a pressure vehicle into the customer's silo. In industrial countries, 80% or more of cement is delivered in bulk.

Typical Constituents of Portland Clinker Plus Gypsum Cement Chemists Notation under CCN.

Clinker	CCN	Mass %
Tricalcium silicate $(CaO)_3 \cdot SiO_2$	C_3S	45–75%
Dicalcium silicate $(CaO)_2 \cdot SiO_2$	C_2S	7–32%
Tricalcium aluminate $(CaO)_3 \cdot Al_2O_3$	C_3A	0–13%
Tetracalcium aluminoferrite $(CaO)_4 \cdot Al_2O_3 \cdot Fe_2O_3$	C_4AF	0–18%
Gypsum $CaSO_4 \cdot 2\ H_2O$		2–10%

Typical Constituents of Portland Cement

Cement Chemists Notation under CCN.

Cement	*CCN*	*Mass %*
Calcium oxide, CaO	C	61–67%
Silicon dioxide, SiO_2	S	19–23%
Aluminum oxide, Al_2O_3	A	2.5–6%
Ferric oxide, Fe_2O_3	F	0–6%
Sulfate	S	1.5–4.5%

Setting and Hardening

Cement sets when mixed with water by way of a complex series of chemical reactions still only partly understood. The different constituents slowly crystallise and the interlocking of their crystals gives cement its strength. Carbon dioxide is slowly absorbed to convert the portlandite ($Ca(OH)_2$) into insoluble calcium carbonate. After the initial setting, immersion in warm water will speed up setting. Gypsum is added as an inhibitor to prevent flash setting.

Use

The most common use for Portland cement is in the production of concrete. Concrete is a composite material consisting of aggregate (gravel and sand), cement, and water. As a construction material, concrete can be cast in almost any shape desired, and once hardened, can become a structural (load bearing) element. Users may be involved in the fas panels, beams, road furniture, or may make cast-*in situ* concrete such as building superstructures, roads, dams. These may be supplied with concrete mixed on site, or may be provided with "ready-mixed" concrete made at permanent mixing sites. Portland cement is also used in mortars (with sand and water only) for plasters and screeds, and in grouts (cement/water mixes squeezed into gaps to consolidate foundations, road-beds, etc.).

When water is mixed with Portland cement, the product sets in a few hours and hardens over a period of weeks. These processes can vary widely depending upon the mix used and the conditions of curing of the product, but a typical concrete sets in about 6 hours and develops a compressive strength of 8 MPa in 24 hours. The strength rises to 15 MPa at 3 days, 23 MPa at 1 week, 35 MPa at 4 weeks and 41 MPa at 3 months. In principle, the strength continues to rise slowly as long as water is available for continued hydration, but concrete is usually allowed to dry out after a few weeks and this causes strength growth to stop.

Types

General

There are different standards for classification of Portland cement. The two major standards are the ASTM C150 used primarily in the USA and European EN 197. EN 197 cement types CEM I, II, III, IV, and V do not correspond to the similarly named cement types in ASTM C150.

ASTM C150

There are five types of Portland cements with variations of the first three according to ASTM C150.

Type I Portland cement is known as common or general purpose cement. It is generally assumed unless another type is specified. It is commonly used for general construction especially when making precast and precast-prestressed concrete that is not to be in contact with soils or ground water. The typical compound compositions of this type are:

55% (C_3S), 19% (C_2S), 10% (C_3A), 7% (C_4AF), 2.8% MgO, 2.9% (SO_3), 1.0% Ignition loss, and 1.0% free CaO.

A limitation on the composition is that the (C_3A) shall not exceed fifteen percent.

Type II gives off less heat during hydration. This type of cement costs about the same as Type I. Its typical compound composition is:

51% (C_3S), 24% (C_2S), 6% (C_3A), 11% (C_4AF), 2.9% MgO, 2.5% (SO_3), 0.8% Ignition loss, and 1.0% free CaO.

A limitation on the composition is that the (C_3A) shall not exceed eight percent which reduces its vulnerability to sulfates. This type is for general construction that is exposed to moderate sulfate attack and is meant for use when concrete is in contact with soils and ground water especially in the western United States due to the high sulfur content of the soil. Because of similar price to that of Type I, Type II is much used as a general purpose cement, and the majority of Portland cement sold in North America meets this specification.

Note: Cement meeting (among others) the specifications for types I and II has become commonly available on the world market.

Type III has relatively high early strength. Its typical compound composition is:

57% (C_3S), 19% (C_2S), 10% (C_3A), 7% (C_4AF), 3.0% MgO, 3.1% (SO_3), 0.9% Ignition loss, and 1.3% free CaO.

This cement is similar to Type I, but ground finer. Some manufacturers make a separate clinker with higher C_3S and/or C_3A content, but this is increasingly rare, and the general purpose clinker is usually used, ground to a specific surface typically 50–80% higher. The gypsum level may also be increased a small amount. This gives the concrete using this type of cement a three day compressive strength equal to the seven day compressive strength of types I and II. Its seven day compressive strength is almost equal to types I and II 28 day compressive strengths.

The only downside is that the six month strength of type III is the same or slightly less than that of types I and II. Therefore the long-term strength is sacrificed a little. It is usually used for precast concrete manufacture, where high 1-day strength allows fast turnover of molds. It may also be used in emergency construction and repairs and construction of machine bases and gate installations.

Type IV Portland cement is generally known for its low heat of hydration. Its typical compound composition is:

28% (C_3S), 49% (C_2S), 4% (C_3A), 12% (C_4AF), 1.8% MgO, 1.9% (SO_3), 0.9% Ignition loss, and 0.8% free CaO.

The percentages of (C_2S) and (C_4AF) are relatively high and (C_3S) and (C_3A) are relatively low. A limitation on this type is that the maximum percentage of (C_3A) is seven, and the maximum percentage of (C_3S) is thirty-five. This causes the heat given off by the hydration reaction to develop at a slower rate. However, as a consequence the strength of the concrete develops slowly. After one or two years the strength is higher than the other types after full curing. This cement is used for very large concrete structures, such as dams, which have a low surface to volume ratio. This type of cement is generally not stocked by manufacturers but some might consider a large special order. This type of cement has not been made for many years, because Portland-pozzolan cements and ground granulated blast furnace slag addition offer a cheaper and more reliable alternative.

Type V is used where sulfate resistance is important. Its typical compound composition is:

38% (C_3S), 43% (C_2S), 4% (C_3A), 9% (C_4AF), 1.9% MgO, 1.8% (SO_3), 0.9% Ignition loss, and 0.8% free CaO.

This cement has a very low (C_3A) composition which accounts for its high sulfate resistance. The maximum content of (C_3A) allowed is five percent for Type V Portland cement. Another limitation is that the (C_4AF) + 2(C_3A) composition cannot exceed twenty percent. This type is used in concrete that is to be exposed to alkali soil and ground water sulfates which react with (C_3A) causing disruptive expansion. It is unavailable in many places although its use is common in the western United States and Canada. As with Type IV, Type V Portland cement has mainly been supplanted by the use of ordinary cement with added ground granulated blast furnace slag or tertiary blended cements containing slag and fly ash.

Types Ia, IIa, and IIIa have the same composition as types I, II, and III. The only difference is that in Ia, IIa, and IIIa an air-entraining agent is ground into the mix. The air-entrainment must meet the minimum and maximum optional specification found in the ASTM manual. These types are only available in the eastern United States and Canada but can only be found on a limited basis. They are a poor approach to air-entrainment which improves resistance to freezing under low temperatures.

Types II(MH) and II(MH)a have a similar composition as types II and IIa but with a mild heat.

EN 197

EN 197-1 defines 5 classes of common cement that comprise Portland cement as a main constituent. These classes differ from the ASTM classes.

I	Portland cement	Comprising Portland cement and up to 5% of minor additional constituents
II	Portland-composite cement	Portland cement and up to 35% of other single constituents
III	Blastfurnace cement	Portland cement and higher percentages of blastfurnace slag
IV	Pozzolanic cement	Portland cement and up to 55% of pozzolanic constituents(volcanic ash)
V	Composite cement	Portland cement, blastfurnace slag or fly ash and pozzolana

Constituents that are permitted in Portland-composite cements are artificial pozzolans (blastfurnace slag, silica fume, and fly ashes) or natural pozzolans (siliceous or siliceous aluminous materials such as volcanic ash glasses, calcined clays and shale).

White Portland Cement

White Portland cement or white ordinary Portland cement (WOPC) is similar to ordinary, grey Portland cement in all respects except for its high degree of whiteness. Obtaining this colour requires some modification to the method of manufacture; because of this, it is somewhat more expensive than the grey product.

The main requirement is to have low iron content which should be less than 0.5% expressed as Fe_2O_3 for white cement and less than 0.9% for off-white cement. It helps to have the iron oxide as ferrous oxide (FeO) which is obtained via slight reducing conditions i.e. operating with zero excess oxygen at the kiln exit. This gives the clinker and cement a green tinge. Other metals such as Cr, Mn, Ti etc. in trace content can also give colour tinges so for a project it is best to use cement from a single source.

Safety Issues

Bags of cement routinely have health and safety warnings printed on them because not only is cement highly alkaline, but the setting process is exothermic. As a result, wet cement is strongly caustic and can easily cause severe skin burns if not promptly washed off with water. Similarly, dry cement powder in contact with mucous membranes can cause severe eye or respiratory irritation. The reaction of cement dust with moisture in the sinuses and lungs can also cause a chemical burn as well as headaches, fatigue, and lung cancer. The production of comparatively low-alkalinity cements (pH<11) is an area of ongoing investigation.

In Scandinavia, France, and UK, the level of chromium(VI), which is considered to be toxic and a major skin irritant, may not exceed 2 ppm.

Environmental Effects

Portland cement manufacture can cause environmental impacts at all stages of the process. These include emissions of airborne pollution in the form of dust, gases, noise, and vibration when operating machinery and during blasting in quarries, consumption of large quantities of fuel during manufacture, release of CO_2 from the raw materials during manufacture, and damage to countryside from quarrying. Equipment to reduce dust emissions during quarrying and manufacture of cement is widely used, and equipment to trap and separate exhaust gases are coming into increased use. Environmental protection also includes the re-integration of quarries into the countryside after they have been closed down by returning them to nature or re-cultivating them.

Epidemiologic Notes and Reports Sulfur Dioxide Exposure in Portland Cement Plants, from the Centres for Disease Control, states "Workers at Portland cement facilities, particularly those burning fuel containing sulfur, should be aware of the acute and chronic effects of exposure to SO_2 [sulfur dioxide], and peak and full-shift concentrations of SO_2 should be periodically measured."

"The Arizona Department of Environmental Quality was informed this week that the Arizona Portland Cement Co. failed a second round of testing for emissions of hazardous air pollutants at the company's Rillito plant near Tucson. The latest round of testing, performed in January 2003 by the company, is designed to ensure that the facility complies with federal standards governing the emissions of dioxins and furans, which are byproducts of the manufacturing process." Cement Reviews' "Environmental News" web page details case after case of environmental problems with cement manufacturing.

An independent research effort of AEA Technology to identify critical issues for the cement industry today concluded the most important environment, health and safety performance issues facing the cement industry are atmospheric releases (including greenhouse gas emissions, dioxin, NO_x, SO_2, and particulates), accidents, and worker exposure to dust.

The CO_2 associated with Portland cement manufacture falls into 3 categories:

Source 1. CO_2 derived from decarbonation of limestone,

Source 2. CO_2 from kiln fuel combustion,

Source 3. CO_2 produced by vehicles in cement plants and distribution.

Source 1 is fairly constant: minimum around 0.47 kg CO_2 per kg of cement, maximum 0.54, typical value around 0.50 worldwide. Source 2 varies with plant efficiency: efficient precalciner plant 0.24 kg CO_2 per kg cement, low-efficiency wet process as high as 0.65, typical modern practices (e.g. UK) averaging around 0.30. Source 3 is almost insignificant at 0.002–0.005. So typical total CO_2 is around 0.80 kg CO_2 per kg finished cement.

This omits the CO_2 associated with electric power consumption because this varies according to the local generation type and efficiency. Typical electrical energy consumption is of the order of 90–150 kWh per tonne cement, equivalent to 0.09–0.15 kg CO_2 per kg finished cement if the electricity is coal-generated.

Overall, with nuclear- or hydroelectric power and efficient manufacturing, CO_2 generation can be reduced to 0.7 kg per kg cement, but can be twice as high. The thrust of innovation for the future is to reduce sources 1 and 2 by modification of the chemistry of cement, by the use of wastes, and by adopting more efficient processes. Although cement manufacturing is clearly a very large CO_2 emitter, concrete (of which cement makes up about 15%) compares quite favourably with other building systems in this regard.

Cement Plants Used for Waste Disposal or Processing=

Due to the high temperatures inside cement kilns, combined with the oxidizing (oxygen-rich) atmosphere and long residence times, cement kilns are used as a processing option for various types of waste streams: indeed, they efficiently destroy many hazardous organic compounds.

The waste streams also often contain combustible materials which allow the substitution of part of the fossil fuel normally used in the process.

Waste materials used in cement kilns as a fuel supplement:

- Car and truck tires – steel belts are easily tolerated in the kilns
- Paint sludge from automobile industries
- Waste solvents and lubricants
- Meat and bone meal – slaughterhouse waste due to bovine spongiform encephalopathy contamination concerns
- Waste plastics
- Sewage sludge
- Rice hulls
- Sugarcane waste
- Used wooden railroad ties (railway sleepers)
- Spent Cell Liner (SCL) from the aluminium smelting industry (also called Spent Pot Liner or SPL)

Portland cement manufacture also has the potential to benefit from using industrial by-products from the waste-stream. These include in particular:

- Slag
- Fly ash (from power plants)
- Silica fume (from steel mills)
- Synthetic gypsum (from desulfurisation)

Nano-Concrete

Concrete is a complex material made of just four basic ingredients: coarse aggregates (stone), fine aggregates (sand), water, and cement. It is extremely hardy and maintenance free, leading to some claims that it is a green material. But cement is incredibly energy intensive to produce. Minerals must be fired in ovens at temperatures of 1500 degrees C (over 2700 degrees F). Because concrete is used so frequently cement manufacture is actually responsible for an estimated 5 to 10 percent of all carbon dioxide released to the atmosphere.

MIT researchers have been examining the structure of concrete in an attempt to better understand how the material works and to develop better (and less energy-intensive methods) for manufacturing cement.

Cutting back on cement production, or finding a cleaner way of producing it could be a huge step in reducing greenhouse gasses in the atmosphere. It may not be as cool as light transmitting concrete, but it has a much greater potential to be a useful contribution towards a greener world.

Nanoengineered Concrete could Cut Carbon Dioxide Emissions

While government leaders argue about the practicality of reducing world emissions of carbon dioxide, scientists and engineers are seeking ways to make it happen.

One group of engineers at MIT decided to focus its work on the nanostructure of concrete, the world's most widely used material. The production of cement, the primary component of concrete, accounts for 5 to 10 percent of the world's total carbon dioxide emissions; the process is an important contributor to global warming.

In the January issue of the Journal of the Mechanics and Physics of Solids, the team reports that the source of concrete's strength and durability lies in the organisation of its nanoparticles. The discovery could one day lead to a major reduction in carbon dioxide emissions during manufacturing.

The work also shows that the study of very common materials at the nano scale has great potential for improving materials in ways we might not have conceived. Ulm refers to this work as the "identification of the geogenomic code of materials, the blueprint of a material's nanomechanical behavior."

Cement is manufactured at the rate of 2.35 billion tons per year, enough to produce 1 cubic metre of concrete for every person in the world. If engineers can reduce carbon dioxide emissions in the world's cement manufacturing by even 10 percent, that would accomplish one-fifth of the Kyoto Protocol goal of a 5.2 percent reduction in total carbon dioxide emissions.

Cement, the oldest engineered construction material, dating back to the Roman Empire, starts out as limestone and clay that are crushed to a powder and heated to a very high temperature (1500 degrees Celsius) in a kiln. At this high temperature, the mineral undergoes a transformation, storing energy in the powder. When the powder is mixed with water, the energy is released into chemical bonds to form the elementary building block of cement, calcium-silicate-hydrate (C-S-H). At the micro level, C-S-H acts as a glue to bind sand and gravel together to create concrete. Most of the carbon dioxide emissions in this manufacturing process result from heating the kiln to a temperature high enough to transfer energy into the powder.

Ulm and Constantinides gathered a wide range of cement pastes from around the world, and, using a novel nano-indentation technique, poked and prodded the hardened cement paste with a nano-sized needle. An atomic force microscope allowed them to see the nanostructure and judge the strength of each paste by measuring indentations created by the needle, a technique that had been used before on homogenous materials, but not on a heterogeneous material like cement.

To their surprise, they discovered that the C-S-H behavior in all of the different cement pastes consistently displays a unique nanosignature, which they call the material's genomic code. This indicates that the strength of cement paste, and thus of concrete, does not lie in the specific mineral, but in the organisation of that mineral as packed nanoparticles.

The C-S-H particles (each about five nanometres, or billionths of a metre, in diameter) have only two packing densities, one for particles placed randomly, say in a box, and another for those stacked symmetrically in a pyramid shape (like a grocer's pile of fruit). These correspond exactly to the mathematically proved highest packing densities allowed by nature for spherical objects: 63 and 74 percent, respectively. In other words, the MIT research shows that materials pack similarly even at the nano scale.

Opportunities & Challenges for Use of Nanotechnology in Cement-Based Materials

Nanotechnology has immense potential and abilities to control the materials world including cement-based materials. It is the science of extremely tiny particle that deals with the study and use of materials and devices that operate at unimaginable nanometre scale i.e. one billionth, or 10"9, of a metre. A nanometre is so small that we cannot see something of this size without the use of very powerful microscopes. It is used to measure things that are very tiny i.e. atoms and molecules. Another way of putting it: a nanometre is the amount an average man's beard grows in the time it takes him to raise the razor to his face. It is more about taking advantage of novel properties that arise solely due to the nano-scale and producing useful or functional components. Materials at this scale behave very differently from when they are in larger form. This technology offers better built, long lasting, cleaner, safer and smarter products for the home, for medicine, for agriculture, for communication and for industry in general. It is the synthesis and application of ideas from science and engineering towards the understanding and production of new materials and devices with enhanced properties and capabilities.

Nanoscience & Nanotechnology

Nanoscience is the study of phenomena and manipulation of materials at atomic, molecular and macromolecular scales, where properties differ significantly from those at a larger scale; while nanotechnology describes design, characterisation, production and application of materials and systems by controlling the size at nanometre scale. The term "Nanotechnology" was invented by Taniguchi in 1974, and it was first popularized in the 1980 by scientist and visionary K. Eric Drexler in his book "Engines of Creation." Nanotechnology is the creation and utilisation of materials, devices, and systems through the control of matter on the nanometre scale. It is more about taking advantage of novel properties that arise solely due to the nano-scale and producing useful or functional components with new or enhanced properties. Some of the important fields where nanotechnology applications are being actively explored are shown in Figure earlier.

Why do Materials at Nano-Scale Behave Differently?

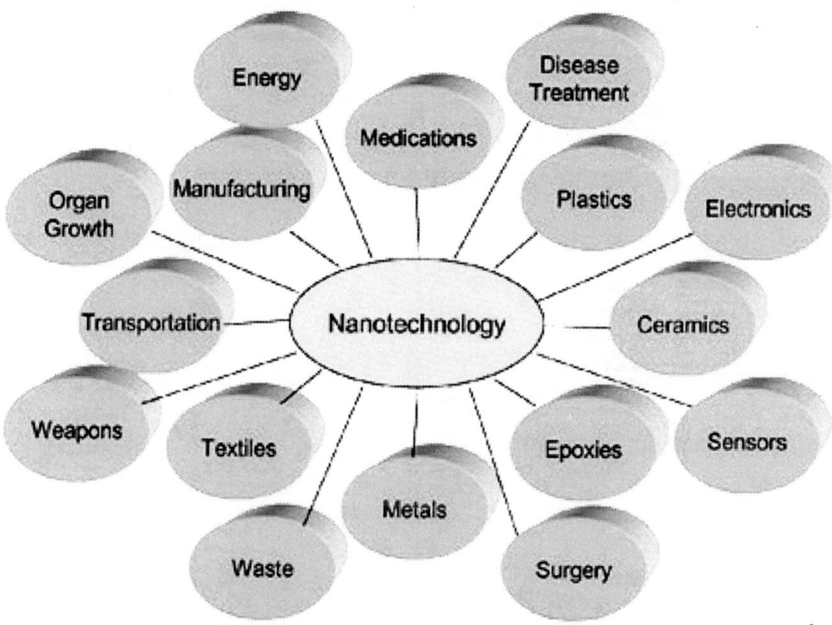

Figure: *Current active fields where nanotechnology applications are being explored*

When the dimensions of a material are decreased from macro size to nano size, significant changes in electronic conductivity, optical absorption, chemical reactivity and mechanical properties occur. Materials with the size range from 100 nm down to the atomic level can have different or enhanced properties compared with the same materials at a larger size. The two main reasons for the change in behavior are an increased relative surface area and the dominance of quantum effects. An increase in surface area will result in a corresponding increase in chemical reactivity and making some nanoparticles useful as catalysts too. As size of the matter is reduced to tens of nanometres or less, quantum effects can begin to play a role, and these can significantly change materials optical, magnetic or electrical properties.

Carbon Nanotubes (CNTs)

Carbon nanotubes are among the most extensively researched nanomaterials today. CNTs are tubular structures of nano metre diameter with large aspect ratio. These tubes have attracted much attention in recent years not only for their small dimensions but also for their potential applications in various fields. A single sheet of graphite is called grapheme. A CNT can be produced by curling a graphite sheet. Carbon sheets can also curl in number of ways. CNT can be considered as the most superior carbon fibre ever made. Addition of small amount (1% by wt) of CNT can improve the mechanical properties consisting of the main Portland cement phase and water. A CNT can be singled or multi walled. CNTs are the strongest and most flexible molecular material with Young's modulus of over 1 TPa. The approximate diameter is 1 nm with length to micron order. CNTs have excellent flexibility. These are essentially free from defects. Nanotubes are highly resistant to chemical attack and have a high strength to weight ratio (1.8g/ cm3 for MWNTs & 0.8G/cm3 for SWNTs). CNT has maximum strain of about 10% which is higher than any other material. Electrical conductivity of CNTs are six orders of magnitude higher than copper, hence, have very high current carrying capacity. Hence, carbon nano tubes have excellence potential for use in the cement composites.

Opportunities in the Fields of Cement-Based Composites

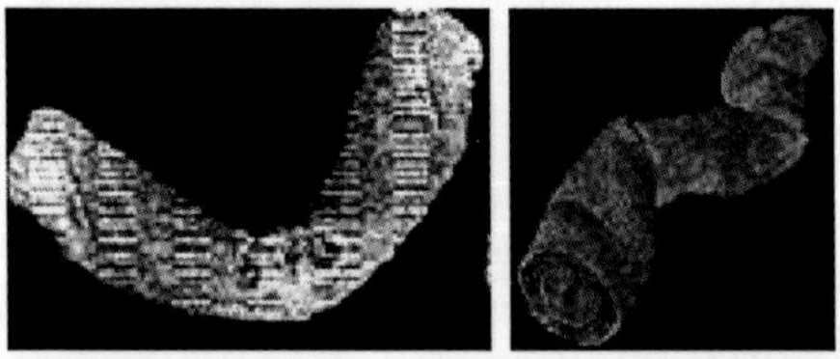

Figure: *Flexible behavior of CNTs*

Nanotechnology is being used for the creation of new materials, devices and systems at molecular, nano- and micro-level. Interest in nanotechnology concept for portland cement composites is steadily growing. The most reported research work regarding application of nanotechnology in cement-based materials is either related to coating or enhancement of mechanical and electrical properties. Some of the widely reported nanoparticles in cement concrete industries are Titanium dioxide (TiO2), Nanosilica (SiO2), Alumina (Al2O3), Carbon nanotube (CNT) etc. Currently, the most active research areas dealing with cement and concrete are: understanding of the hydration of cement particles and the use of nano-size ingredients such as alumina and silica particles. A typical scale of various constituent of a normal concrete is given in Figure earlier.

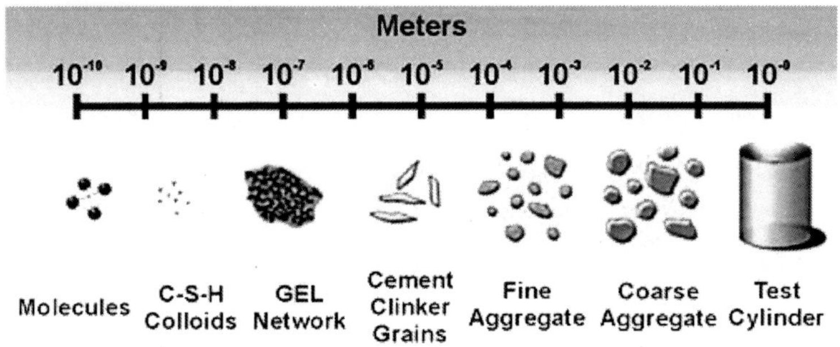

Figure: *Scales of various constituents of concrete*

Average size of Portland cement particle is about 50 microns. In applications that require thinner final products and faster setting time, micro cement with a maximum particle size of about 5 microns is being used. Knowledge at the nanoscale of the structure and characteristics of materials will promote the development of new applications and new products to repair or improve the properties of construction materials. For example, the structure of the fundamental calcium-silicate-hydrate (C-S-H) gel which is responsible for the mechanical and physical properties of cement pastes, including shrinkage, creep, porosity, permeability and elasticity. C-S-H gel can be modified to obtain better durability. Cement-based materials containing carbon nanotubes can be used for both strengthening and enhancing electrical and electronic properties of the concrete besides their mechanical properties. Development of smart concrete using carbon nano tubes would be easier. If nano-cement particles can be processed with nanotubes and nano-size silica particles; conductive, strong, tough, more flexible, cement-based composites can be developed with enhanced properties, for electronic applications and coatings.

Nano Concrete and Nano Ingredients

Nano concrete is defined as a concrete made with portland cement particles with sizes ranging from a few nanometre to a maximum of about 100 micrometres. Nano ingredients are ingredients with at least one dimension of nano metre size. Therefore the particle size has to be reduced in order to obtain nano-portland cement. If these nano-cement particles can be processed with nanotubes and reactive nano-size silica particles; conductive, strong, tough, more flexible, cement-based composites can be developed with enhanced properties, for electronic applications and coatings. There is also limited information dealing with the manufacture of nano-cement. If cement with nano-size particles can be manufactured and processed, it will open up a large number of opportunities in the fields of cement-based composites. Current research activity in concrete using nano cement and nano silica includes:

- Characterization of cement hydration
- Influence of the addition of nano-size silica to concrete
- Synthesis of cement using nano particles and coatings (applied to protect concrete).

The characteristics of nano concrete can be effectively used to create unique products, which can be also molded to complex shapes. Nano concrete containing carbon nanotubes can be used for both strengthening and creating electric circuits. Recently, Wang et al. has demonstrated the use of nanoclays in the making of self compacting concrete for improve it for slipform paving for the construction of concrete road.

The NIST researchers have claimed that the service life of concrete could be doubled by using a nano-sized additive that slows down aggressive ion transport in concrete. Rather than change the size and density of the pores in concrete, they reasoned, it would be better to change the viscosity of the solution in the concrete to reduce the speed at which chlorides and sulfates enter the concrete.

The basic is "Swimming through a pool of honey takes longer than making it through a pool of water. The NIST researchers have demonstrated that the nanotechnology additives can be blended directly into the concrete with current chemical admixtures, but that even better performance is achieved when the additives are mixed into the concrete by saturating absorbent, lightweight sand.

It has been claimed that all the samples showed a unique nanosignature for the C-S-H material regardless of the sample origin. This is known as the material's genomic code and means that the strength of cement paste is not a function of specific minerals but relates to the way the nanoparticles are arranged.

"If everything depends on the organisational structure of the nanoparticles that make up concrete, rather than on the material itself, we can conceivably replace it with a material that has concrete's other characteristics—strength, durability, mass availability and low cost—but does not release so much CO2 into the atmosphere during manufacture" said Professor Franz-Josef Ulm from civil and environmental engineering at MIT.

Nano Silica Fume for Improving Concrete Performance

Nano silica is most common nano additive to concrete. It is reported that nano silica was found to be much effective than micron sized silica for improving the performance such as permeability, and subsequently, durability. In addition, reduced amount of about 15 to 20 kg of nano silica was found to provide same strength as 60 kg of regular or micro silica.

Nano-silica is effective additives to polymers and concrete, a development realised in high-performance and self–compacting concrete with improved workability and strength. Nano-silica addition to cement based materials control the degradation of the fundamental C-S-H (Calcium-silicate-hydrate) reaction in water as well as block water penetration and leads to improvement in durability. The addition of nano SiO2 particles enhances the density and strength of concrete. The results indicate that nanosilica behaves not only as a filler to improve microstructure, but also as an activator to promote pozzolanic reaction for fly ash concrete as a result strength of the fly ash concrete improves particularly in the early stages.

Coatings for Concrete

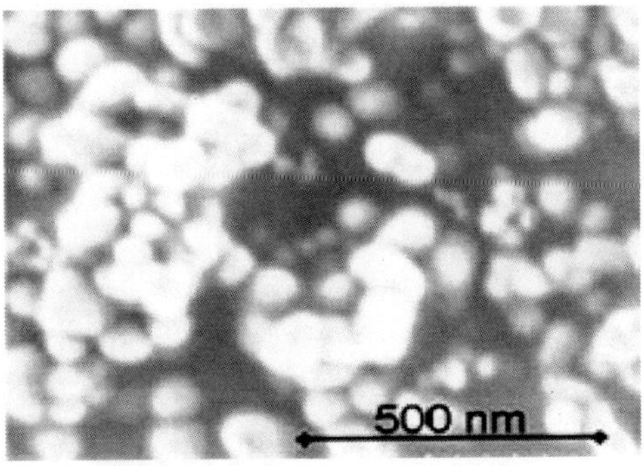

Figure: *SEM of a typical nanosilica particles*

Another major large volume application of nano powder in cement-based materials is the area of coatings. The attractive colouring on ancient Czech glasses is found to contain nanoparticles. This shows that nanotechnology was used for coating surfaces i.e. spraying and making a product look attractive from ancient time. Nano-powders have a remarkable surface area. The surface area imparts a serious change of surface energy and surface morphology. The change in properties causes improved catalytic ability, tunable wavelength-sensing ability and better designed pigments and paints with self-cleaning and self-healing feature. One promising area of application of nanoparticle for cement based materials is development of self-cleaning coating.

Titanium oxide is commonly used for this purpose. It is incorporated, as nano particles to block UV light. It is added to paints, cements and windows for its sterilizing properties as TiO2 breaks down organic pollutants, volatile organic compounds, and bacterial membrane through powerful catalyst reactions and can reduce airborne pollutants applying to outer surfaces. Additionally it is hydrophilic and therefore gives self-cleaning properties to surface to which it is applied.

Challenges for the Application of Nanotechnology in Cement-Based Materials

The primary challenge is to manufacture nano-size cement particles. The second challenge is the heat of hydration. Due to many folds increase in specific surface areas the nano cement particle are very reactive and results in enormous amount of heat of hydration. Special organic and inorganic additives are needed to be developed to control the setting and heat of hydration. For the nano-coatings, the properties of the coatings themselves need extensive investigation. Durability of coatings under various exposure conditions, abrasion resistance, friction resistance, high temperature resistance, failure modes, and electrical characteristics need to be proven by using exhaustive experiments. Furthermore, huge amount of expenditure is needed for the equipments to study the behaviour of nano cement and nano additive in cement based materials.

Nanoconcrete: Smart Multifunctional Cementitious Nanocomposite Material

The nanomaterials give the material a smart structure with inherent sensor capabilities whereby its structural health can be monitored.

The material is made by adding a nano admixture to the water used in a conventional cementitious material manufacturing process. The cement concrete-carbon nanotube structure displays high tensile and flexural strengths and fracture toughness, low electrical resistivity, and high thermal conductivity and electromagnetic interference shielding effectiveness.

Concrete is an essential material for building all types of infrastructure including buildings, roads, dams, etc. As for brittle materials in general, concrete is strong under compression and weak under tension or flexure and has the tendency to crack and has poor thermal conductivity. The exothermic reaction in the formation of concrete and the poor thermal conductivity of the materials in concrete pose a problem for large pours such as dams, where coolant systems must be built into the slabs for proper curing.

The nanoconcrete manufacturing method uses a process to create nano admixtures that transform traditional cementitious composite material, such as concrete, into multifunctional smart structural material (nanoconcrete) without adding any additional weight and without altering the manufacturing process. The process and materials are easy to scale up. Also, the nano admixture introduces ductility, durability and other multifunctionalities without sacrificing existing concrete properties.

The Nanoconcrete uses a nanomaterials based admixture for use in reinforcements and multifunctional performance of cementitious composites. The nanomaterials in the admixture may be graphitic or non-graphitic carbon materials or nanoparticles such as silicon carbide, dispersed and stabilized in water. The carbon material that is preferably used in the nanoconcrete is carbon nanotube. The carbon nanotubes used may be single walled, multiple walled, as prepared or functionalized.

The nano admixture-reinforced structural materials, such as cementitious composites, have smart materials characteristics and multifunctionalities. In contrast to other approaches, the nanoconcrete admixture does not alter the manufacturing process, is easy to incorporate in traditional concrete materials and to scale-up, and is cost effective. The newly developed nano admixture transforms traditional concrete material into multifunctional smart nanoconcrete without adding any additional weight.

The presence of nano-admixture in concrete will also result in significant improvements in crack resistance, thermal conductivity, and fibre dispersion.

The many advantages of Nanoconcrete include: lower environmental impact—advanced concrete will result in higher-strength, less-volume, lighter-weight structures with lower CO2 emissions; simplicity of use—builders/workers and other users will require no special training to manufacture NanoConcrete and safety—nanotubes and nanoparticles will be suspended and stabilized in a solution, mitigating safety risks for users.

Cement and aggregates are provided and mixed together. The nano admixture modified water is mixed with the mixture of cement and aggregates to form a concrete. The concrete is placed in a mold, compacted, and cured. Chemical additives or fibres may be provided, and may be mixed with and at the same time as the cement and aggregates. Silica fume may be used as a dispersing or densifying agent.

The cementitious nanocomposite exhibits improved mechanical and chemical properties without significant additional weight compared to a corresponding conventional or carbon fibre reinforced cementitious material without nanomaterials.

The material displays high tensile and flexural strengths, fracture toughness, thermal conductivity, and electromagnetic interference shielding effectiveness and low electrical resistivity compared to corresponding conventional or carbon fibre reinforced cementitious material without nanomaterials.

The material exhibits significantly improved crack resistance, vibration damping capacity, permeability, air void content, steel rebar corrosion resistance, coefficient of thermal expansion, thermal conductivity, fibre dispersion, and workability compared to corresponding conventional or carbon fibre reinforced cementitious material without nanomaterials.

The material has improved alkali silica reaction inhibition properties compared to corresponding conventional or carbon fibre reinforced cementitious material without nanomaterials. The nanomaterials reduce the permeability of the material, preventing the ingress of water. The nanomaterials strengthen the interfacial transition zones in the material, acting as tiny mechanical rebar to suppress the effects of alkali silica reaction.

The sensor capabilities of the material include a change in volume electrical response depending on applied stress, whereby strain on the material can be detected by measuring its electrical resistance. Roads are paved with such a material, the material's electrical resistance at different points is measured, and the measurements are used to determine the location, weight, and speed of traffic.

The nanomaterials may comprise carbon nanotubes and reduce pores and prevent micro-cracking by fibre-bridging across micro crack regions. The carbon nanotubes form a stitching on fracture surfaces, diverting crack energy into a matrix and inhibiting crack propagation. The negative coefficient of thermal expansion of the carbon nanotubes results in a higher thermal stability for the material.

Exploring the Micromechanics of Openended Pile Driving via Discrete Element

A pile driver is a mechanical device used to drive piles (poles) into soil to provide foundation support for buildings or other structures. The term is also used in reference to members of the construction crew that work with pile-driving rigs.

One traditional type of pile driver includes a heavy weight placed between guides so that it is able to freely slide up and down in a single line. It is placed upon a pile. The weight is raised, which may involve the use of hydraulics, steam, diesel, or manual labour. When the weight reaches its highest point it is then released and smashes on to the pile in order to drive it into the gr. There are a number of claims to the invention of the pile driver. A mechanically sound drawing of a pile driver appeared as early as 1475 in Francesco di Giorgio Martini's treatise *Trattato di Architectura.* Also, several other prominent inventors — James Nasmyth (son of Alexander Nasmyth), watchmaker James Valoué, Count Giovan Battista Gazzola, and Leonardo da Vinci — have all been credited with inventing the device. However, there is evidence that a comparable device was used in the construction of Crannogs at Oakbank and Loch Tay in Scotland as early as 5000 years ago. Otis Tufts is credited with inventing the steam pile driver in the United States.

Types

Ancient pile driving equipment used manual or animal labour to lift heavy weights, usually by means of pulleys, to drop the weight onto the end of the pile. Modern piledriving equipment uses various methods to raise the weight and guide the pile.

Diesel Hammer

A modern diesel pile hammer is a very large two-stroke diesel engine. The weight is the piston, and the apparatus which connects to the top of the pile is the cylinder. Piledriving is started by having the weight raised by auxiliary means — usually a cable from the crane holding the pile driver — which draws air into the cylinder. Diesel fuel is added/injected into the cylinder. The weight is dropped, using a quick-release. The weight of the piston compresses the air/fuel mixture, heating it to the ignition point of diesel fuel. The mixture ignites, transferring the energy of the falling weight to the pile head, and driving the weight back up. The rising weight draws in fresh air, and the cycle starts over until the fuel runs out or is stopped by the pile crew.

From an army manual on pile driving hammers: The initial start up of the hammer requires the piston (ram) to be raised to a point where the trip automatically releases the piston, allowing it to fall by gravity. As the piston falls, it activates the fuel pump, which discharges a metred amount of fuel into the ball pan of the impact block. The falling piston also blocks the exhaust ports, and compression of fuel trapped in the cylinder begins. The compressed air exerts a pre-load force to hold the impact block firmly against the drive cap and pile. At the bottom of the compression stroke, the piston strikes the impact block, atomizing the fuel and starting the pile on its downward movement. In the instant after the piston strikes, the atomized fuel ignites, and the resulting explosion exerts an even greater force on the already moving pile, driving it further into the ground. The reaction of the explosion rebounding from the resistance of the pile drives the piston upward. As the piston rises, the exhaust ports open, releasing the gases and force of the explosion into the atmosphere. After the piston stops its upward movement, it again falls by gravity to start another cycle.

Vertical Travel Lead Systems

The vertical travel lead, referred to as "VTL" system, was first developed and patented by C.W. Bermingham in the 1970s. This lead system was developed in response to the fundamental limitations of either a fixed lead or swinging lead system. The fixed lead system is well suited to level job sites with few obstructions and has the advantage of faster positioning of the lead. The hanging lead is very adaptable to different elevations and batter piles but takes much longer to position. The Vertical Travel Lead was developed to combine the fast and accurate positioning of fixed leads, with the ability to adjust the height of the lead base up or down. The VTL lead is connected to the boom by a sliding connection, which allows the lead to be elevated or lowered below grade. The VTL system has become the industry standard in Canada, US Railway Construction, and many parts of the USA. Vertical Travel Leads come in two main forms: Spud and Box Lead types. Box leads are very common in the Southern United States and Spud Leads are common in the Northern United States, Canada and Europe.

Hydraulic Hammer

A hydraulic hammer is a modern type of piling hammer used in place of diesel and air hammers for driving steel pipe, precast concrete, and timber piles. Hydraulic hammers are more environmentally acceptable than the older, less efficient hammers as they generate less noise and pollutants. However, in many cases the dominant noise is caused by the impact of the hammer on the pile, or the impacts between components of the hammer, so that the resulting noise level can be very similar to diesel hammers.

Hydraulic Press-in

Speciality equipment which installs piles using hydraulic rams to press piles into the ground. This system is preferred where vibration is a concern. There are press attachments that can adapt to conventional pile driving rigs to press 2 pairs of sheet piles at a time. Additional types of press equipment sit on top of existing sheet piles and grip onto previously driven piles. This system allows for greater press-in and extraction force to be used since more reaction force is developed. The reaction based machines operate at only 69dB at 23 ft allowing for installation and extraction of piles in very close proximity to noise and vibration sensitive areas where traditional methods may threaten the stability of existing structures.

Such equipment and methods are specified into portions of the internal drainage system in the New Orleans area after Hurricane Katrina as well as many projects around the world where noise, vibrations and limited access are a concern during the engineering, design and construction phases of the project.

Vibratory Pile Driver/Extractor

Vibratory pile hammers contain a system of counter-rotating eccentric weights, powered by hydraulic motors, and designed in such a way that horizontal vibrations cancel out, while vertical vibrations are transmitted into the pile. The pile driving machine is lifted and positioned over the pile by means of an excavator or crane, and is fastened to the pile by a clamp and/or bolts. Vibratory hammers can either drive in or extract a pile; extraction is commonly used to recover steel "H" piles used in temporary foundation shoring. Hydraulic fluid is typically supplied to the driver by a diesel engine powered pump mounted in a trailer or van and connected to the driver head through a set of long hoses. When the pile driver is connected to a dragline excavator, it is powered by the excavator's own diesel engine. Vibratory pile drivers are often chosen to mitigate noise, as when the construction is very close to residence or office buildings, or when there is not enough vertical clearance above the foundation to permit use of a conventional pile hammer (for example when retrofitting additional piles to a bridge column or abutment footing). Hammers are available with several different vibration rates, ranging from about 1200 vibrations per minute to about 2400 VPM; the vibration rate chosen is influenced by soil conditions at the site and other factors such as power requirements and purchase price of the equipment.

Piling Rig

A piling rig is a construction machine for piling in foundation engineering. It is mainly applied to drill in sandy soil, clay, silty clay, etc. and widely used cast-in-place piles, diaphragm walls, foundation reinforcement and other foundation projects. Its rated power of engine is around 108–450 kW, output torque 60–400 kN•m, maximum pile diameter 1.5–4 m, maximum pile depth 60–90 m. It can meet construction requirement of kinds of foundation engineering projects. It generally applies hydraulic crawler chassis, automatic lift box-type mast, telescopic drill pipe, auto vertical adjustor, depth indicator, etc. Its operation applies hydraulic pilot control, load sensor, thus it is easy and comfortable to control. Main and auxiliary hoisting can meet different requirements in construction sites. The rig can be used on piling construction in dry (short screw), wet soil (rotary bucket) and rock (core drill) by equipping with drilling tool, and can be equipped with long screw drill, diaphragm wall grab, vibratory hammer, etc., to realise many functions. It is mainly used in foundation engineering of municipal construction, expressway, bridge, industrial and civil buildings, diaphragm wall, water conservancy project, Slope protection, etc.

Piling Rigs Categories

1. small-sized – torque is around 60–100 kN•m, engine power 108 kW, drilling diameter 0.5–1.2m, drilling depth 40m, total quality 40t.
2. middle-sized – torque is around 120–180 kN•m, engine power 125–200 kW, drilling diameter 0.8–1.8m, drilling depth 60m, total quality 42–65t.
3. large-sized - torque is around 240 kN•m, engine power 300 kW, drilling diameter 1–2.5m, drilling depth 80m, total quality 100t

Environmental Effects

The underwater sound pressure caused by pile-driving may be deleterious to fish when conducted adjacent to fish-bearing waters. State and local regulatory agencies manage environment issues associated with pile-driving.

7

Synthesis and Characterization of Nanoparticulate Calcium Aluminates

Calcium aluminates are a range of minerals obtained by heating calcium oxide and aluminium oxide together at high temperatures. They are encountered in the manufacture of refractories and cements.

The stable phases shown in the phase diagram (formed at atmospheric pressure under an atmosphere of normal humidity) are:

- Tricalcium aluminate, $3CaO \cdot Al_2O_3$ (C3A)
- Dodecacalcium hepta-aluminate, $12CaO \cdot 7Al_2O_3$ (C12A7)
- Monocalcium aluminate, $CaO \cdot Al_2O_3$ (CA) (mayenite)
- Monocalcium dialuminate, $CaO \cdot 2Al_2O_3$ (CA2)
- Monocalcium hexa-aluminate, $CaO \cdot 6Al_2O_3$ (CA6)

In addition, other phases include:

- Dicalcium aluminate, $2CaO \cdot Al_2O_3$ (C2A), which exists only at pressures above 2500 MPa. The crystal is orthorhombic, with density 3480 kg·m"3.
- Pentacalcium trialuminate, $5CaO \cdot 3Al_2O_3$ (C5A3), forms only under an anhydrous and oxygen free atmosphere. The crystal is orthorhombic, with a density of 3067 kg·m"3. It reacts rapidly with water.
- Tetracalcium trialuminate, $4CaO \cdot 3Al_2O_3$ (C4A3), is a metastable phase formed by dehydrating $4CaO \cdot 3Al_2O_3 \cdot 3H_2O$ (C4A3H3).

Calcium Aluminate Cements

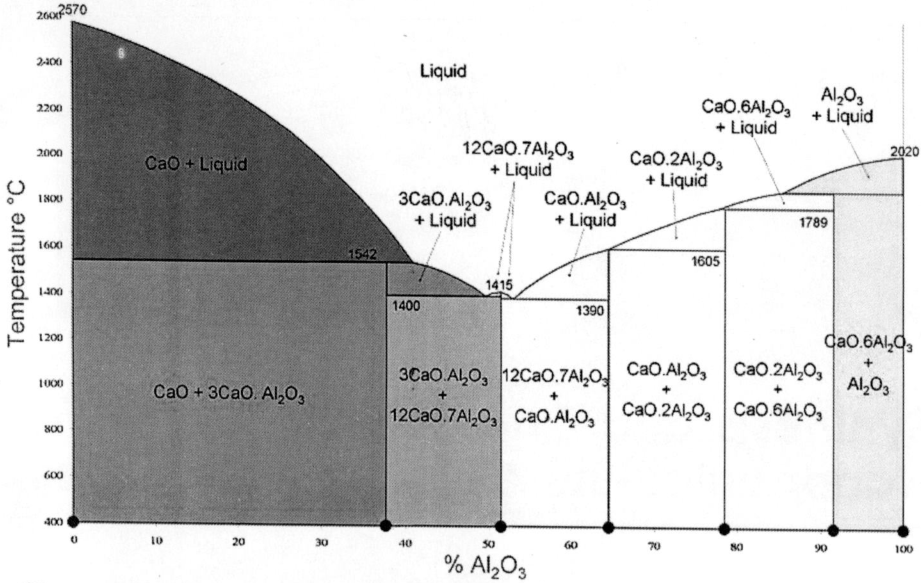

***Figure:** Phase diagram of calcium aluminates present in the anhydrous calcium aluminate cement before hydration.*

Calcium aluminate cements are cements consisting predominantly of hydraulic calcium aluminates. Alternative names are "aluminous cement", "high-alumina cement" and "Ciment fondu" in French. They are used in a number of small-scale, specialised applications.

The method of making cement from limestone and low-silica bauxite was patented in France in 1908 by Bied of the Pavin de Lafarge Company. The initial development was as a result of the search for a cement offering sulfate resistance.

The cement was called in the French "Ciment fondu". Subsequently, its other special properties were discovered, and these guaranteed its future in niche applications.

Composition

The main active constituent of calcium aluminate cements is monocalcium aluminate ($CaAl_2O_4$, $CaO \cdot Al_2O_3$, or CA in the cement chemist notation).

It usually contains other calcium aluminates as well as a number of less reactive phases deriving from impurities in the raw materials. Rather a wide range of compositions is encountered, depending on the application and the purity of aluminium source used. Constituents of some typical formulations include:

Oxide/Mineral	*General Purpose*	*Buff*	*White*	*Refractory*
SiO_2	4.0	5.0	2.7	0.4
Al_2O_3	39.4	53.0	62.4	79.6
Fe_2O_3	16.4	2.0	0.4	0
CaO	38.4	38.0	34.0	19.8
MgO	1.0	0.1	0.1	0
Na_2O	0.1	0.1	0	0
K_2O	0.2	0	0	0
TiO_2	1.9	1.8	0.4	0.1
Monocalcium aluminate	46	70	70	35
Dodecacalcium hepta-aluminate	10	5	0	0
Monocalcium dialuminate	0	0	17	30
Belite	7	5	0	0
Gehlenite	4	14	11	1
Ferrite	24	5	2	0
Pleocroite	1	1	1	0
Wüstite	7	0	0	0
Corundum	0	0	0	33

The mineral phases all take the form of solid solutions with somewhat variable compositions

Manufacture

The cement is made by fusing together a mixture of a calcium-bearing material (normally calcium oxide from limestone) and an aluminium-bearing material (normally bauxite for general purposes, or refined alumina for white and refractory cements). The liquified mixture cools to a basalt-like clinker which is ground alone to produce the finished product. Because complete melting usually takes place, raw materials in lump-form can be used.

A typical kiln arrangement comprises a reverberatory furnace provided with a shaft preheater in which the hot exhaust gases pass upward as the lump raw material mix passes downward. The preheater recuperates most of the heat in the combustion gases, dehydrates and de-hydroxylates the bauxite and de-carbonates the limestone. The calcined material drops into the "cool end" of the melt bath. The melt overflows the hot end of the furnace into molds in which it cools and solidifies. The system is fired with pulverized coal or oil. The cooled clinker ingots are crushed and ground in a ball-mill. In the case of high-alumina refractory cements, where the mix only sinters, a rotary kiln can be used.

Reaction with Water

The hydration reactions of calcium aluminate cements are very complex. The strength-developing phases are monocalcium aluminate, dodeca-calcium hepta-aluminate and belite. Calcium aluminoferrite, monocalcium dialuminate, gehlenite and pleochroite contribute little to strength.

The reactive aluminates react with water initially to form a mixture of: $CaO \cdot Al_2O_3 \cdot 10\ H_2O$,$_2$ $CaO \cdot Al_2O_3 \cdot 8\ H_2O$,$_3$ $CaO \cdot Al_2O_3 \cdot 6\ H_2O$, and $Al(OH)_3$ gel, the amounts of each depending upon the curing temperature. The first two hydrates subsequently decompose to a mixture of $3\ CaO \cdot Al_2O_3 \cdot 6\ H_2O$, $Al(OH)_3$ gel and water, this process being called "conversion". Because of the loss of water, conversion causes an increase in porosity, which can be accompanied by a decrease in strength. This need not be a problem in structural concrete provided that a sufficiently high cement content and a sufficiently low water/cement ratio are employed.

Applications

Because of their relatively high cost, calcium aluminate cements are used in a number of restricted applications:

- in construction concretes, rapid strength development is achieved, even at low temperatures.
- in construction concretes, high chemical resistance is possible.
- in refractory concretes, strength is maintained at high temperatures.
- as a component in blended cement formulations, various properties such as ultra-rapid strength development and controlled expansion can be obtained.

Problems

Incorrect use of calcium aluminate cements has led to widespread construction problems, especially during the third quarter of the 20th century when this type of cement was used because of its faster hardening properties. After several years some of the buildings and structures collapsed due to degradation of the cement and many had to be torn down or reinforced. Heat and humidity accelerate the degradation process called "conversion"; the roof of a swimming pool was one of the first structures to collapse in the UK.

In Madrid, Spain, a large housing block nicknamed Korea (because it was built to house Americans during the Korean War), built 1951~1954 was affected and had to be torn down in 2006. Also in Madrid the Vicente Calderón soccer stadium was affected and had to be partially rebuilt and reinforced.

Effects of Watercement Ratio and Curing Age on the Threshold Pore Width

The water–cement ratio is the ratio of the weight of water to the weight of cement used in a concrete mix and has an important influence on the quality of concrete produced. A lower water-cement ratio leads to higher strength and durability, but may make the mix more difficult to place. Placement difficulties can be resolved by using plasticizers or super-plasticizers.

Often, the water–cement ratio is characterized as the water to cement plus pozzolan ratio, w/(c+p). The pozzolan is typically a fly ash, or blast furnace slag. It can include a number of other materials, such as silica fume, rice hull ash or natural pozzolans. The addition of pozzolans will influence the strength gain of the concrete.

The concept of water–cement ratio was developed by Duff A. Abrams and first published in 1918.

Concrete hardens as a result of the chemical reaction between cement and water (known as hydration, this produces heat and is called the heat of hydration). For every pound (or kilogram or any unit of weight) of cement, about 0.25 pounds (or 0.25 kg or corresponding unit) of water is needed to fully complete the hydration reactions. This requires a water-cement ratio of 1:4 often given as a proportion: 0.25. However, a mix with a w/c ratio of 0.25 may not mix thoroughly, and may not flow well enough to be placed, so more water is used than is technically necessary to react with the cement. More typical water-cement ratios of 0.4 to 0.6 are used. For higher-strength concrete, lower water:cement ratios are used, along with a plasticizer to increase flowability.

Too much water will result in segregation of the sand and aggregate components from the cement paste. Also, water that is not consumed by the hydration reaction may leave the concrete as it hardens, resulting in microscopic pores(bleeding) that will reduce the final strength of the concrete. A mix with too much water will experience more shrinkage as the excess water leaves, resulting in internal cracks and visible fractures (particularly around inside corners) which again will reduce the final strength.

The 1997 Uniform Building Code specifies a maximum 0.50 water-to-cement ratio (1:2) when concrete is exposed to freezing and thawing in a moist condition or to de-icing chemicals, and a maximum 0.45 water to cement ratio for concrete in severe or very severe sulfate conditions.

Nanostructure of Single Carbon Fibres Investigated with Synchrotron Radiation

Carbon fibre, alternatively graphite fibre, carbon graphite or CF, is a material consisting of fibres about 5–10 μm in diameter and composed mostly of carbon atoms.

To produce carbon fibre, the carbon atoms are bonded together in crystals that are more or less aligned parallel to the long axis of the fibre as the crystal alignment gives the fibre high strength-to-volume ratio (making it strong for its size). Several thousand carbon fibres are bundled together to form a tow, which may be used by itself or woven into a fabric.

The properties of carbon fibres, such as high stiffness, high tensile strength, low weight, high chemical resistance, high temperature tolerance and low thermal expansion, make them very popular in aerospace, civil engineering, military, and motorsports, along with other competition sports. However, they are relatively expensive when compared to similar fibres, such as glass fibres or plastic fibres.

Carbon fibres are usually combined with other materials to form a composite. When combined with a plastic resin and wound or molded it forms carbon fibre reinforced polymer (often referred to as carbon fibre) which has a very high strength-to-weight ratio, and is extremely rigid although somewhat brittle. However, carbon fibres are also composed with other materials, such as with graphite to form carbon-carbon composites, which have a very high heat tolerance.

In 1879, Thomas Edison baked cotton threads or bamboo slivers at high temperatures carbonizing them into an all-carbon fibre filament used in the first incandescent light bulb to be heated by electricity. In 1958, Roger Bacon created high-performance carbon fibres at the Union Carbide Parma Technical Centre, now GrafTech International Holdings, Inc., located outside of Cleveland, Ohio. Those fibres were manufactured by heating strands of rayon until they carbonized. This process proved to be inefficient, as the resulting fibres contained only about 20% carbon and had low strength and stiffness properties. In the early 1960s, a process was developed by Dr. Akio Shindo at Agency of Industrial Science and Technology of Japan, using polyacrylonitrile (PAN) as a raw material. This had produced a carbon fibre that contained about 55% carbon.

The high potential strength of carbon fibre was realised in 1963 in a process developed by W. Watt, L. N. Phillips, and W. Johnson at the Royal Aircraft Establishment at Farnborough, Hampshire. The process was patented by the UK Ministry of Defence, then licensed by the NRDC to three British companies: Rolls-Royce already making carbon fibre, Morganite, and Courtaulds. Within a few years, Rolls-Royce took advantage of the new material's properties to broke into the American market with its RB-211 aero-engine with carbon fibre's compressor blades. Unfortunately, the blades proved vulnerable to damage from bird impact. That and other Rolls-Royce's problems became so great that the company was nationalized in 1971 and the carbon-fibre production plant was sold off to form "Bristol Composites".

In the late 1960s, the Japanese took the lead in manufacturing PAN-based carbon fibres. The 1970 joint technology agreement allowed Union Carbide to manufacture the Japan's Toray Industries superior product and USA to dominate the market. Morganite decided that carbon-fibre production was peripheral to its core business, leaving Courtaulds as the only big UK manufacturer. Continuing collaboration with the staff at Farnborough proved helpful in the quest for higher quality and improvements in the speed of production as Courtaulds developed two main markets: aerospacc and sports equipment. However Courtaulds's big advantage as manufacturer of the "Courtelle" precursor now became a weakness. Courtelle's low cost and ready availability were potential advantages, but the water-based inorganic process used to produce it made the product susceptible to impurities that did not affect the organic process used by other carbon-fibre manufacturers.

Nevertheless, during the 1980s Courtaulds continued to be a major supplier of carbon fibre for the sports-goods market, with Mitsubishi its main customer until a move to expand, including building a production plant in California, turned out badly. The investment did not generate the anticipated returns, leading to a decision to pull out of the area and Courtaulds ceased carbon-fibre production in 1991. Ironically the one surviving UK carbon-fibre manufacturer continued to thrive making fibre based on Courtaulds's precursor. Inverness-based RK Carbon Fibres Ltd concentrated on producing carbon fibre for industrial applications, removing the need to compete at the quality levels reached by overseas manufacturers.

During the 1960s, experimental work to find alternative raw materials led to the introduction of carbon fibres made from a petroleum pitch derived from oil processing. These fibres contained about 85% carbon and had excellent flexural strength. Also, during this period, the Japanese Government heavily supporting carbon fibre development at home and several Japanese companies such as Toray, Nippon Carbon, Toho Rayon and Mitsubishi started their own development and production. As they subsequently advanced to become market leaders, companies in USA and Europe were encouraged to take up these activities as well, either through their own developments or contractual acquisition of carbon fibre knowledge. These companies included Hercules, BASF and Celanese USA and Akzo in Europe.

Since the late 1970s, further types of carbon fibre yarn entered the global market, offering higher tensile strength and higher elastic modulus. For example, T400 from Toray with a tensile strength of 4,000 MPa and M40, a modulus of 400 GPa. Intermediate carbon fibres, such as IM 600 from Toho Reyon with up to 6,000 MPa were developed. Carbon fibres from Toray, Celanese and Akzo found their way to aerospace application from secondary to primary parts first in military and later in civil aircraft as in McDonnell Douglas, Boeing and Airbus planes. By 2000 the industrial applications for highly sophisticated machine parts in middle Europe was becoming more important.

Structure and Properties

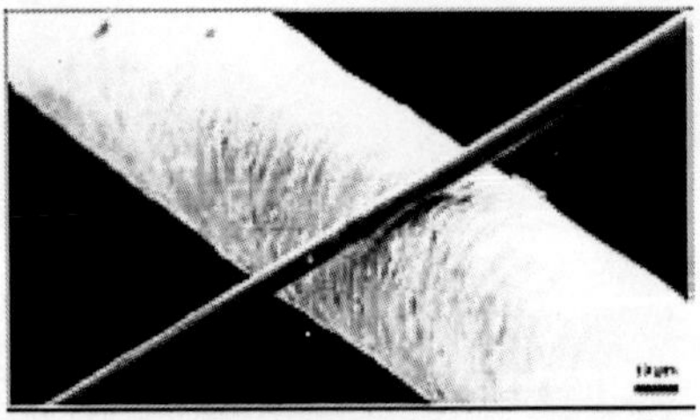

Figure: *A 6 μm diameter carbon filament (running from bottom left to top right) compared to a human hair.*

Each carbon filament thread is a bundle of many thousand carbon filaments. A single such filament is a thin tube with a diameter of 5–8 micrometres and consists almost exclusively of carbon. The earliest generation (e.g. T300, HTA and AS4) had diameters of 16–22 micrometres. Later fibres (e.g. IM6 or IM600) have diameters that are approximately 5 micrometres.

The atomic structure of carbon fibre is similar to that of graphite, consisting of sheets of carbon atoms (graphene sheets) arranged in a regular hexagonal pattern, the difference being in the way these sheets interlock. Graphite is a crystalline material in which the sheets are stacked parallel to one another in regular fashion. The intermolecular forces between the sheets are relatively weak Van der Waals forces, giving graphite its soft and brittle characteristics.

Depending upon the precursor to make the fibre, carbon fibre may be turbostratic or graphitic, or have a hybrid structure with both graphitic and turbostratic parts present.

In turbostratic carbon fibre the sheets of carbon atoms are haphazardly folded, or crumpled, together. Carbon fibres derived from Polyacrylonitrile (PAN) are turbostratic, whereas carbon fibres derived from mesophase pitch are graphitic after heat treatment at temperatures exceeding 2200 °C.

Turbostratic carbon fibres tend to have high tensile strength, whereas heat-treated mesophase-pitch-derived carbon fibres have high Young's modulus (i.e., high stiffness or resistance to extension under load) and high thermal conductivity.

Applications

Figure: *Tail of an RC helicopter, made of Carbon fibre reinforced polymer*

The global demand on carbon fibre composites was valued at roughly US$10.8 billion in 2009, which declined 8–10% from the previous year. It is expected to reach US$13.2 billion by 2012 and to increase to US$18.6 billion by 2015 with an annual growth rate of 7% or more. Strongest demands come from aircraft & aerospace, wind energy, as well as from the automotive industry with optimized resin systems.

Composite Materials

Carbon fibre is most notably used to reinforce composite materials, particularly the class of materials known as carbon fibre or graphite reinforced polymers. Non-polymer materials can also be used as the matrix for carbon fibres.

Due to the formation of metal carbides and corrosion considerations, carbon has seen limited success in metal matrix composite applications. Reinforced carbon-carbon (RCC) consists of carbon fibre-reinforced graphite, and is used structurally in high-temperature applications. The fibre also finds use in filtration of high-temperature gases, as an electrode with high surface area and impeccable corrosion resistance, and as an anti-static component.

Molding a thin layer of carbon fibres significantly improves fire resistance of polymers or thermoset composites because a dense, compact layer of carbon fibres efficiently reflects heat.

The increasing use of carbon fibre composites is displacing aluminum from aerospace applications in favour of other metals because of galvanic corrosion issues.

Textiles

Precursors for carbon fibres are polyacrylonitrile (PAN), rayon and pitch. Carbon fibre filament yarns are used in several processing techniques: the direct uses are for prepregging, filament winding, pultrusion, weaving, braiding, etc. Carbon fibre yarn is rated by the linear density (weight per unit length, i.e. 1 g/1000 m = 1 tex) or by number of filaments per yarn count, in thousands. For example, 200 tex for 3,000 filaments of carbon fibre is three times as strong as 1,000 carbon filament yarn, but is also three times as heavy. This thread can then be used to weave a carbon fibre filament fabric or cloth. The appearance of this fabric generally depends on the linear density of the yarn and the weave chosen. Some commonly used types of weave are twill, satin and plain. Carbon filament yarns can be also knitted or braided.

Microelectrodes

Carbon fibres are used for fabrication of carbon-fibre microelectrodes. In this application typically a single carbon fibre with diameter of 5–7 μm is sealed in a glass capillary. At the tip the capillary is either sealed with epoxy and polished to make carbon-fibre disk microelectrode or the fibre is cut to a length of 75–150 μm to make carbon-fibre cylinder electrode. Carbon-fibre microelectrodes are used either in amperometry or fast-scan cyclic voltammetry for detection of biochemical signaling.

Catalysis

PAN-based nanofibres can efficiently catalyze the first step in the making of synthetic gasoline (syngas) and other energy-rich products out of carbon dioxide. The process uses a "co-catalyst" system in three steps:

(1) EMIM–CO2 complex formation;

(2) adsorption of EMIM–CO2 complex on reduced carbon atoms and

(3) Carbon monoxide formation.

The first step uses an ionic liquid, while graphitic carbon structures doped with other reactive atoms replaced silver to produce the final output. The carbon nanofibre catalyst exhibited negligible overpotential (0.17 V) for carbon dioxide reduction and more than an order of magnitude higher current density compared with silver under similar experimental conditions. The reduction derived from the reduced carbons rather than to electronegative nitrogen dopants. The performance came from the nanofibrillar structure and high binding energy of key intermediates to the carbon nanofibre surfaces.

Synthesis

Each carbon filament is produced from a precursor polymer such as polyacrylonitrile (PAN), rayon, or petroleum pitch. For synthetic polymers such as PAN or rayon, the precursor is first spun into filament yarns, using chemical and mechanical processes to initially align the polymer atoms in a way to enhance the final physical properties of the completed carbon fibre. Precursor compositions and mechanical processes used during spinning filament yarns may vary among manufacturers. After drawing or spinning, the polymer filament yarns are then heated to drive off non-carbon atoms (carbonization), producing the final carbon fibre. The carbon fibres filament yarns may be further treated to improve handling qualities, then wound on to bobbins.

A common method of manufacture involves heating the spun PAN filaments to approximately 300 °C in air, which breaks many of the hydrogen bonds and oxidizes the material. The oxidized PAN is then placed into a furnace having an inert atmosphere of a gas such as argon, and heated to approximately 2000 °C, which induces graphitization of the material, changing the molecular bond structure. When heated in the correct conditions, these chains bond side-to-side (ladder polymers), forming narrow graphene sheets which eventually merge to form a single, columnar filament. The result is usually 93–95% carbon. Lower-quality fibre can be manufactured using pitch or rayon as the precursor instead of PAN. The carbon can become further enhanced, as high modulus, or high strength carbon, by heat treatment processes. Carbon heated in the range of 1500–2000 °C (carbonization) exhibits the highest tensile strength (820,000 psi, 5,650 MPa or N/mm^2), while carbon fibre heated from 2500 to 3000 °C (graphitizing) exhibits a higher modulus of elasticity (77,000,000 psi or 531 GPa or 531 kN/mm^2).

Manufacturers

Major manufacturers of carbon fibres include Cytec Industries, Formosa Plastics, Hexcel, Mitsubishi Rayon, SGL Carbon, Toho Tenax, Toray Industries and Zoltek. Manufacturers typically make different grades of fibres for different applications. Higher modulus carbon fibres are typically more expensive.

Concrete Petrography

Petrographic analysis is an important type of analysis when evaluating the quality and durability of concrete. Petrographic analysis is usually a naturally part of e.g. a failure analysis of concrete and may be performed at different scales, starting with macro-examination using the naked eye and stereomicroscope, micro-examination using the optical polarizing microscope and nano-examination using the scanning electron microscope (SEM).

All methods reveal at their own level information about the concrete composition and structure. Petrographic analysis is a fast and reliable technique which gives answers to questions such as:

- Why does my concrete perform like this?
- Why does it deteriorate?
- Was the concrete correctly handled during proportioning, mixing and casting?
- What is the expected durability of my concrete?

Having this in mind petrographic analysis is the ideal method to use in research, quality control, mix design optimization and forensic examination.

Petrographic testing is the use of microscopes to examine samples of rock or concrete to determine their mineralogical and chemical characteristics. Samples for petrographic examination can be taken from lump samples or cores. They are impregnated with resin. Suitable surfaces (either polished or thin sections) are prepared by sawing, lapping and polishing. The samples are examined through a petrological (geological polarising) microscope, using either reflected or transmitted light. The lighting may be normal, ultra-violet or polarised.

Petrographic examination can determine a range of parameters, as follows:

- Type, proportions, grading, shape and condition of the aggregates.
- Nature of the cement paste, including mineral additions such as fly ash/pfa, the degree of hydration.
- Presence of deleterious material, cement paste and air voids.
- Depth of carbonation.
- Bond between the aggregate and the paste.
- Evidence of sulfate attack, frost damage, some forms of chemical attack and alkali-silica reactions.
- Depth of fire damage and temperature of the fire.
- Measurement of air entrainment.

Energetically Modified Cement

Energetically modified cements (EMC) are a class of cementitious materials made from pozzolans (e.g. fly ash, volcanic ash, pozzolana), silica sand, blast furnace slag, or Portland cement (or blends of these ingredients).

Classification and Field-Usage Potential

An energetically modified cement is a cementitious material that has been produced using the EMC Activation process. The term "energetically modified cement" (abbreviated as "EMC" or "EMC cement") refers to a distinct class of cementitious materials. There are several different energetically modified cements depending on the raw materials used.

Although the term "energetically modified cement" implies that such compounds are cements, are more accurately be described as "cementitious materials". EMC cannot fully replace conventional Portland cement in concrete unless Portland cement itself is the raw material undergoing EMC Activation. Where raw materials other than Portland cement undergo EMC Activation, the resultant energetically modified cements are called "Alternative Cementitious Materials" or "Supplemental Cementitious Materials". Colloquially, energetically modified cements not made from Portland cement are sometimes describes as "Green Cements", because of the significant energy and carbon dioxide savings. As such, EMC may be viewed as a contributor to the emerging field of ecodesign.

The usefulness of energetically modified cements depends on the performance characteristics required, based on the mechanical loads expected and the ambient environment. The most useful EMCs are those made from fly ash and natural pozzolans — on account of their relative abundance, the performance characteristics of the respective EMC, the relatively high Portland cement replacement ratios made available by EMC Activation using these raw materials, together with the associated energy and carbon dioxide savings.

EMC products have been extensively tested by independent labs, including Caltrans and other concrete producers.

History

The term "energetically modified cement" is widely accepted in the academic community. The term was first used in Sweden, where the EMC Activation process was discovered in 1992 by Vladimir Ronin at Luleå University of Technology (LTU). The process was refined there by Dr. Ronin and others, including Lennart Elfgren (now Professor Emeritus of LTU, Division of Structural Engineering, Department of Civil, Mining and Environmental Engineering).

The term "energetically modified cement" was used first in a paper by Ronin et al. in 1993. At the 45th World Exhibition of Invention, Research and Innovation, held in Brussels, Belgium, EMC Activation was awarded a Gold Medal with mention by EUREKA, the European inter-governmental (research and development) organisation.

Given that the EMC Activation process is entirely mechanical in nature (as opposed to thermal), its potential to cause significant energy savings has been further recognised independently for a number of years. This recognition continues.

Continuing academic work and research with energetically modified cements is ongoing at LTU, including work within the auspices of the *Sveriges Bygguniversitet* (SBU). The nascent "self-healing" properties of EMCs have some resonance within the emerging field of biomimetics in the advanced material sciences and civil engineering disciplines. In March 2013 Elfgren presented LTU's perspective at the Future Infrastructure Forum (FIF) held at University of Cambridge.

The research work connected with EMCs has received numerous awards from the *Elsa τ Sven Thysells stiftelse fφr konstruktionsteknisk forskning* (Elsa & Sven Thysell Foundation for Construction Engineering Research) of Sweden.

Effect of EMCs on a Concrete's Chemistry and "Self-Healing"

Using pozzolans in concrete provides a number of chemical pathways whereby porous (reactive) Portlandite is transformed into a number of hard and impermeable (relatively non-reactive) compounds, rather than producing the porous and soft relatively reactive calcium carbonate produced using ordinary concrete. Many of the end products of pozzolanic chemistry exhibit a hardness greater than 7.0 on the Mohs scale. By comparison, Tungsten is 7.5 on the scale.

The greater the replacement in the concrete of Portland cement with pozzolanic cementitious materials (of which EMCs are an example), the greater the propensity for the foregoing. EMC Activation is a process which is thought to increase a pozzolan's chemical affinity for such pozzolanic reactions. This is yields a faster and greater strength development of the resulting concrete—at higher replacement ratios—than untreated pozzolans. As such, EMCs may be classified also as "highly reactive pozzolans". Highly reactive pozzolans are thought to yield further stabilisation benefits upon the pozzolanic reaction-pathways.

"Self-Healing" (Autogenous Effects)

It is for the foregoing reasons (and others) that it is thought pozzolanic mortars and concretes have been observed to "self-heal". This effect is considered a natural autogenous property. By virtue that EMC Activation is a process which is thought to increase a pozzolan's affinity for such pozzolanic reactions, concretes made from EMCs are no different in this regard. The same autogenous tendency been noted and studied in the various supporting structures of Hagia Sophia built for the Byzantine emperor Justinian (now, Istanbul, Turkey). There, in common with most Roman cements, mortars comprising high amounts of pozzolana were used — in order to give what is thought to be an increased resistance to the stress-effects caused by the various earthquakes that have disrupted the region throughout the millennia.

Range of Concretes Produced

The performance of concretes made from energetically modified cements can be custom-designed. Hence, concretes can range from those exhibiting superior strength and durability that reduce the carbon footprint at up to ~70% as compared to concretes made from Portland cement, through to the production of rapid and ultra-rapid hardening, high-strength concretes (for example, over 70 MPa / 10,150 psi in 24 hours and over 200 MPa / 29,000 psi in 28 days). This allows energetically modified cements to yield High Performance Concretes (HPCs).

Portland Cement Replacement-Capabilities (and Limitation Considerations)

Generally, the strength and strength-development of pozzolan concretes depend upon the "pozzolanic" characteristics of the raw material that is employed to make it. For example, fly ash in its natural state is typically more "pozzolanic" than volcanic ash — although care should be taken not to necessarily imply that all fly ashes are per se more "pozzolanic" than all volcanic ashes.

In a similar vein, the nascent characteristics of the raw material undergoing EMC Activation may also act as a consideration as to the upper limit of Portland cement replacement by an energetically modified cement.

Moreover, in practical "everyday" terms, the key consideration is a concrete's strength-development within a specified time period. In a project environment, this means a concrete will need to develop a strength within a given time period that either matches or exceeds a project's specifications.

For this reason, although the replacement-ratio of an EMC made from fly ash may exceed even 70%, by comparison the current upper-limit using EMC made from natural pozzolans is 60% for practical large-scale usage.

No Noxious Emissions and Low Leachability of EMCs

The EMC activation of fly ash is entirely mechanical in nature and does not involve any heating or burning in the process. Leachability tests were performed by LTU in 2001 in Sweden on behalf of a Swedish power production company. These tests confirmed that EMC made from fly ash "showed a low surface specific leachability" with respect to "all environmentally relevant metals."

EMCs Using Californian Volcanic Ash and Significance

Energetically modified cements have been used in large infrastructure projects in the United States. When EMC is made from fly ash, high Portland cement replacements (i.e., the replacement of at least 50% Portland cement) yield concretes consistent field results in high-volume applications. This is also the case for EMC made from natural pozzolans (e.g., volcanic ash).

For example, volcanic ash deposits from Southern California of the United States were independently tested. At 50% Portland cement replacement, the resulting concretes exceeded requirements:

- At 28 days, the compressive strength was 4,180 psi / 28.8 MPa (N/mm^2).
- The 56-day strength exceeded the requirements for 4,500 psi (31.1 MPa) concrete, even taking into account the safety margin as recommended by the American Concrete Institute.

The results demonstrated that:

- EMC Activation "has a sufficient positive impact on the water requirement to obtain satisfactory workability and strength of concrete, at about 50% replacement of Portland cement."
- the "index of pozzolanic activity at 7 days was 80% and at 28 days was 88%, which exceeded the relevant standard's requirements (75% at both ages)."

The particle-size distribution and morphology of the EMC produced were studied by Luleå University of Technology. Those studies "evidenced the improvement in the surface smoothness of particles of natural pozzolans processed by the proprietary EMC method."

Durability of Concretes Produced and High Performance Concretes (HPCs)

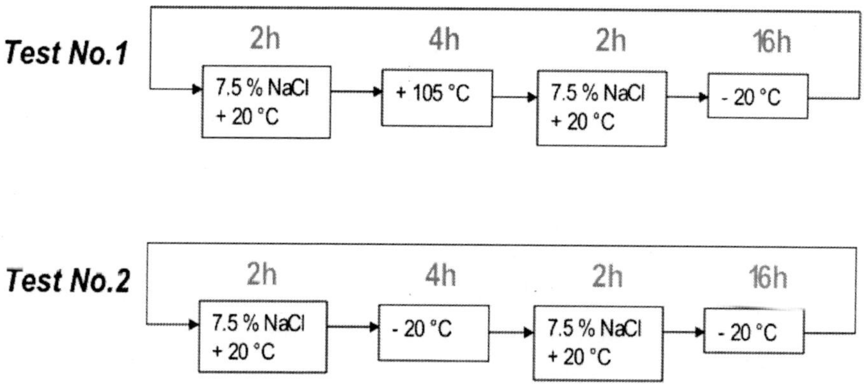

Concrete Durability 2-Test Cycle per Bache (1983)

Diagram: "Bache method" for testing concrete durability, which simulates daily temperature variations in brine. Test 1 or 2 (each 24 h) may be used, or both performed sequentially during a 48-hour period. The chosen cycle is repeated ad nausem in order to determine the mass-loss, as an analogue for durability.

Treating Portland cement with EMC Activation will yield High Performance Concretes. These HPCs will be high strength, highly durable, and exhibiting greater strength-development in contrast to HPCs made from untreated Portland cement, which can have moderate to challenging durability impairments by comparison.

For example, durability tests were been performed according to the "Bache method". The Bache method induces the sequence of saturation by salt water of 7.5% sodium chloride (i.e., a brine, which by definition is of greater salt concentration than sea waters), followed by freezing or heating in a 24-hour cycle, in order to simulate high diurnal temperature ranges. Concrete made from ordinary Portland cement without additives, has a relatively impaired resistance to salt waters. Hence, the Bache method is generally accepted as one of the most severe testing procedures for concrete.

Samples made of high performance concrete comprising (a) EMC (comprising Portland cement and silica fume both having undergone EMC Activation) and (b) Portland cement, having respective compressive strengths of 180.3 and 128.4 MPa (26,150 and 18,622 psi) after 28 days of curing, were then tested using the Bache method. The resulting mass-loss was plotted in order to determine durability. The test results showed:

- EMC high performance concrete showed a "consistent high-level durability" throughout the entire testing-period. For example, "practically no scaling of the concrete was been observed", even after 80 Bache method cycles.
- Whereas, the reference Portland cement concrete had undergone "total destruction after about 16 Bache method cycles, in line with Bache's own observations for high-strength concrete."

In other words, treating Portland cement with the EMC Activation process, may increase the strength-development by nearly 50% and also significantly improve the durability, as measured according to generally accepted methods. All energetically modified cements also exhibit high resistances to chloride and sulphate ion attack, together with low Alkali-Silica Reactivities (ASR). These features allow concretes made from energetically modified cements to exhibit superior durabilities as compared to concretes made from Portland cement, a feature common to all concretes comprising pozzolans. An early project using EMC made from fly ash was the construction of a road bridge in Karungi, Sweden, with Swedish construction firm Skanska. The Karungi road bridge has successfully withstood the tests of time, despite Karungi's harsh subarctic climate and extremely divergent annual and diurnal temperature ranges.

Projects Using EMC

In the United States, energetically modified cements have been approved for usage by a number of State wide Department of Transport agencies, including PennDOT, TxDOT and CalTrans. In the United States, highway bridges and hundreds of miles of highway paving have been constructed using concretes made from EMC derived from fly ash. These projects include the paving of sections of Interstate 10, which is the main U.S. Interstate highway linking Miami, Florida with Los Angeles, California. In these projects, EMC replaced at least 50% of the Portland cement in the concrete poured. This is roughly 2.5 times more than the Portland cement replacement typically offered by untreated fly ash per se. In all projects the 28-day strength requirements were exceeded. For example, independent test-data records the highest 28-day strength at 8015 psi (55.26 MPa) against a project requirement of 4400 psi (30.34 MPa).

Another notable project is the extension of the passenger terminals at the Port of Houston, Texas. This project fully exploits energetically modified cement's ability to yield concretes that exhibit high-resistances to chloride– and sulphate–ion permeability (i.e., increased resistance to sea waters), as compared to concretes made from Portland cement.

Production

As of 2010 the 18 years cumulative volume of concrete produced contain a least partially energetically modified cement was over 4,500,000 cu yd (3,440,496 m^3). This 18 year total represents approximately 0.13% of the yearly worldwide concrete production at 3,500,000,000 cubic yards.

Modification of Cement Paste with Silica Fumea NMR Study

Silica fume, also known as microsilica, (CAS number 69012-64-2, EINECS number 273-761-1) is an amorphous (non-crystalline) polymorph of silicon dioxide, silica. It is an ultrafine powder collected as a by-product of the silicon and ferrosilicon alloy production and consists of spherical particles with an average particle diameter of 150 nm. The main field of application is as pozzolanic material for high performance concrete. It is sometimes confused with fumed silica (also known as pyrogenic silica, CAS number 112945-52-5, EINECS number 231-545-4). However, the production process, particle characteristics and fields of application of fumed silica are all different from those of silica fume.

The first testing of silica fume in Portland-cement-based concretes was carried out in 1952. The biggest drawback to exploring the properties of silica fume was a lack of material to experiment with. Early research used an expensive additive called fumed silica, an amorphous form of silica made by combustion of silicon tetrachloride in a hydrogen-oxygen flame. Silica fume on the other hand, is a very fine pozzolanic, amorphous material, a by-product of the production of elemental silicon or ferrosilicon alloys in electric arc furnaces. Before the late 1960s in Europe and the mid-1970s in the United States, silica fumes were simply vented into the atmosphere. With the implementation of tougher environmental laws during the mid-1970s, silicon smelters began to collect the silica fume and search for its applications. The early work done in Norway received most of the attention, since it had shown that Portland cement-based-concretes containing silica fumes had very high strengths and low porosities. Since then the research and development of silica fume made it one of the world's most valuable and versatile admixtures for concrete and cementitous products.

Production

Silica fume is a byproduct in the carbothermic reduction of high-purity quartz with carbonaceous materials like coal, coke, wood-chips, in electric arc furnaces in the production of silicon and ferrosilicon alloys.

Applications

Concrete: Because of its extreme fineness and high silica content, silica fume is a very effective pozzolanic material. Standard specifications for silica fume used in cementitious mixtures are ASTM C1240, EN 13263. Silica fume is added to Portland cement concrete to improve its properties, in particular its compressive strength, bond strength, and abrasion resistance. These improvements stem from both the mechanical improvements resulting from addition of a very fine powder to the cement paste mix as well as from the pozzolanic reactions between the silica fume and free calcium hydroxide in the paste.

Addition of silica fume also reduces the permeability of concrete to chloride ions, which protects the reinforcing steel of concrete from corrosion, especially in chloride-rich environments such as coastal regions and those of humid continental roadways and runways (because of the use of deicing salts) and saltwater bridges. Prior to the mid-1970s, nearly all silica fume was discharged into the atmosphere. After environmental concerns necessitated the collection and landfilling of silica fume, it became economically viable to use silica fume in various applications, in particular high-performance concrete. Effects of silica fume on different properties of fresh and hardened concrete include

a) *Workability:* With the addition of silica fume, the slump loss with time is directly proportional to increase in the silica fume content due to the introduction of large surface area in the concrete mix by its addition. Although the slump decreases, the mix remains highly cohesive.

b) *Segregation and bleeding:* Silica fume reduces bleeding significantly because the free water is consumed in wetting of the large surface area of the silica fume and hence the free water left in the mix for bleeding also decreases. Silica fume also blocks the pores in the fresh concrete so water within the concrete is not allowed to come to the surface.

8

Fly Ash as Construction Material

Fly ash, also known as flue-ash, is one of the residues generated in combustion, and comprises the fine particles that rise with the flue gases. Ash which does not rise is termed bottom ash. In an industrial context, fly ash usually refers to ash produced during combustion of coal. Fly ash is generally captured by electrostatic precipitators or other particle filtration equipment before the flue gases reach the chimneys of coal-fired power plants, and together with bottom ash removed from the bottom of the furnace is in this case jointly known as coal ash. Depending upon the source and makeup of the coal being burned, the components of fly ash vary considerably, but all fly ash includes substantial amounts of silicon dioxide (SiO_2) (both amorphous and crystalline) and calcium oxide (CaO), both being endemic ingredients in many coal-bearing rock strata.

Toxic constituents depend upon the specific coal bed makeup, but may include one or more of the following elements or substances found in trace quantities (up to hundreds ppm): arsenic, beryllium, boron, cadmium, chromium, hexavalent chromium, cobalt, lead, manganese, mercury, molybdenum, selenium, strontium, thallium, and vanadium, along with dioxins and PAH compounds.

In the past, fly ash was generally released into the atmosphere, but pollution control equipment mandated in recent decades now require that it be captured prior to release. In the US, fly ash is generally stored at coal power plants or placed in landfills. About 43% is recycled, often used as a pozzolan to produce hydraulic cement or hydraulic plaster or a partial replacement for Portland cement in concrete production. Pozzolans ensure the setting of concrete and plaster and provide concrete with more protection from wet conditions and chemical attack.

Some have expressed health concerns about this.

In some cases, such as the burning of solid waste to create electricity ("resource recovery" facilities a.k.a. waste-to-energy facilities), the fly ash may contain higher levels of contaminants than the bottom ash and mixing the fly and bottom ash together brings the proportional levels of contaminants within the range to qualify as nonhazardous waste in a given state, whereas, unmixed, the fly ash would be within the range to qualify as hazardous waste.

Chemical Composition and Classification

Component	*Bituminous*	*Subbituminous*	*Lignite*
SiO_2 (%)	20-60	40-60	15-45
Al_2O_3 (%)	5-35	20-30	20-25
Fe_2O_3 (%)	10-40	4-10	4-15
CaO (%)	1-12	5-30	15-40
LOI (%)	0-15	0-3	0-5

Fly ash material solidifies while suspended in the exhaust gases and is collected by electrostatic precipitators or filter bags. Since the particles solidify rapidly while suspended in the exhaust gases, fly ash particles are generally spherical in shape and range in size from 0.5 μm to 300 μm. The major consequence of the rapid cooling is that only few minerals will have time to crystallize and that mainly amorphous, quenched glass remains. Nevertheless, some refractory phases in the pulverized coal will not melt (entirely) and remain crystalline. In consequence, fly ash is a heterogeneous material. SiO_2, Al_2O_3, Fe_2O_3 and occasionally CaO are the main chemical components present in fly ashes. The mineralogy of fly ashes is very diverse. The main phases encountered are a glass phase, together with quartz, mullite and the iron oxides hematite, magnetite and/or maghemite. Other phases often identified are cristobalite, anhydrite, free lime, periclase, calcite, sylvite, halite, portlandite, rutile and anatase. The Ca-bearing minerals anorthite, gehlenite, akermanite and various calcium silicates and calcium aluminates identical to those found in Portland cement can be identified in Ca-rich fly ashes. The mercury content can reach1 ppm , but is generally included in the range 0.01 - 1 ppm for bituminous coal. The concentrations of other trace elements vary as well according to the kind of coal combusted to form it. In fact, in the case of bituminous coal, with the notable exception of boron, trace element concentrations are generally similar to trace element concentrations in unpolluted soils.

Two classes of fly ash are defined by ASTM C618: Class F fly ash and Class C fly ash. The chief difference between these classes is the amount of calcium, silica, alumina, and iron content in the ash. The chemical properties of the fly ash are largely influenced by the chemical content of the coal burned (i.e., anthracite, bituminous, and lignite).

Not all fly ashes meet ASTM C618 requirements, although depending on the application, this may not be necessary. Ash used as a cement replacement must meet strict construction standards, but no standard environmental regulations have been established in the United States. 75% of the ash must have a fineness of 45 μm or less, and have a carbon content, measured by the loss on ignition (LOI), of less than 4%. In the U.S., LOI needs to be under 6%. The particle size distribution of raw fly ash is very often fluctuating constantly, due to changing performance of the coal mills and the boiler performance. This makes it necessary that, if fly ash is used in an optimal way to replace cement in concrete production, it needs to be processed using beneficiation methods like mechanical air classification. But if fly ash is used also as a filler to replace sand in concrete production, unbeneficiated fly ash with higher LOI can be also used. Especially important is the ongoing quality verification. This is mainly expressed by quality control seals like the Bureau of Indian Standards mark or the DCL mark of the Dubai Municipality.

Class F Fly Ash

The burning of harder, older anthracite and bituminous coal typically produces Class F fly ash. This fly ash is pozzolanic in nature, and contains less than 20% lime (CaO). Possessing pozzolanic properties, the glassy silica and alumina of Class F fly ash requires a cementing agent, such as Portland cement, quicklime, or hydrated lime, with the presence of water in order to react and produce cementitious compounds. Alternatively, the addition of a chemical activator such as sodium silicate (water glass) to a Class F ash can lead to the formation of a geopolymer.

Class C Fly Ash

Fly ash produced from the burning of younger lignite or subbituminous coal, in addition to having pozzolanic properties, also has some self-cementing properties. In the presence of water, Class C fly ash will harden and gain strength over time. Class C fly ash generally contains more than 20% lime (CaO). Unlike Class F, self-cementing Class C fly ash does not require an activator. Alkali and sulfate (SO_4) contents are generally higher in Class C fly ashes.

At least one US manufacturer has announced a fly ash brick containing up to 50% Class C fly ash. Testing shows the bricks meet or exceed the performance standards listed in ASTM C 216 for conventional clay brick; it is also within the allowable shrinkage limits for concrete brick in ASTM C 55, Standard Specification for Concrete Building Brick. It is estimated that the production method used in fly ash bricks will reduce the embodied energy of masonry construction by up to 90%. Bricks and pavers were expected to be available in commercial quantities before the end of 2009.

Disposal and Market Sources

In the past, fly ash produced from coal combustion was simply entrained in flue gases and dispersed into the atmosphere. This created environmental and health concerns that prompted laws which have reduced fly ash emissions to less than 1% of ash produced. Worldwide, more than 65% of fly ash produced from coal power stations is disposed of in landfills and ash ponds.

The recycling of fly ash has become an increasing concern in recent years due to increasing landfill costs and current interest in sustainable development. As of 2005, U.S. coal-fired power plants reported producing 71.1 million tons of fly ash, of which 29.1 million tons were reused in various applications. If the nearly 42 million tons of unused fly ash had been recycled, it would have reduced the need for approximately 27,500 acre·ft (33,900,000 m^3) of landfill space. Other environmental benefits to recycling fly ash includes reducing the demand for virgin materials that would need quarrying and substituting for materials that may be energy-intensive to create such as Portland cement.

As of 2006, about 125 million tons of coal-combustion byproducts, including fly ash, were produced in the U.S. each year, with about 43% of that amount used in commercial applications, according to the American Coal Ash Association Web site. As of early 2008, the United States Environmental Protection Agency hoped that figure would increase to 50% as of 2011.

Fly Ash Reuse

There is no U.S. governmental registration or labelling of fly ash utilisation in the different sectors of the economy - industry, infrastructures and agriculture. Fly ash utilisation survey data, acknowledged as incomplete, are published annually by the American Coal Ash Association.

The ways of coal ash utilisation include (approximately in order of decreasing importance):

- Concrete production, as a substitute material for Portland cement and sand
- Embankments and other structural fills (usually for road construction)
- Grout and Flowable fill production
- Waste stabilization and solidification
- Cement clinkers production - (as a substitute material for clay)
- Mine reclamation
- Stabilization of soft soils
- Road subbase construction
- As Aggregate substitute material (e.g. for brick production)
- Mineral filler in asphaltic concrete
- Agricultural uses: soil amendment, fertilizer, cattle feeders, soil stabilization in stock feed yards, and agricultural stakes
- Loose application on rivers to melt ice
- Loose application on roads and parking lots for ice control
- Other applications include cosmetics, toothpaste, kitchen counter tops, floor and ceiling tiles, bowling balls, flotation devices, stucco, utensils, tool handles, picture frames, auto bodies and boat hulls, cellular concrete, geopolymers, roofing tiles, roofing granules, decking, fireplace mantles, cinder block, PVC pipe, Structural Insulated Panels, house siding and trim, running tracks, blasting grit, recycled plastic lumber, utility poles and crossarms, railway sleepers, highway sound barriers, marine pilings, doors, window frames, scaffolding, sign posts, crypts, columns, railroad ties, vinyl flooring, paving stones, shower stalls, garage doors, park benches, landscape timbers, planters, pallet blocks, molding, mail boxes, artificial reef, binding agent, paints and undercoatings, metal castings, and filler in wood and plastic products.

Portland Cement

Owing to its pozzolanic properties, fly ash is used as a replacement for some of the Portland cement content of concrete. The use of fly ash as a pozzolanic ingredient was recognised as early as 1914, although the earliest noteworthy study of its use was in 1937. Roman structures such as aqueducts or the Pantheon in Rome used volcanic ash or pozzolana (which possesses similar properties to fly ash) as pozzolan in their concrete. As pozzolan greatly improves the strength and durability of concrete, the use of ash is a key factor in their preservation.

Use of fly ash as a partial replacement for Portland cement is particularly suitable but not limited to Class C fly ashes. Class "F" fly ashes can have volatile effects on the entrained air content of concrete, causing reduced resistance to freeze/thaw damage. Fly ash often replaces up to 30% by mass of Portland cement, but can be used in higher dosages in certain applications. Fly ash can add to the concrete's final strength and increase its chemical resistance and durability.

Fly ash can significantly improve the workability of concrete. Recently, techniques have been developed to replace partial cement with high-volume fly ash (50% cement replacement). For roller-compacted concrete (RCC)[used in dam construction], replacement values of 70% have been achieved with processed fly ash at the Ghatghar dam project in Maharashtra, India. Due to the spherical shape of fly ash particles, it can increase workability of cement while reducing water demand. Proponents of fly ash claim that replacing Portland cement with fly ash reduces the greenhouse gas "footprint" of concrete, as the production of one ton of Portland cement generates approximately one ton of CO_2, compared to no CO_2 generated with fly ash. New fly ash production, i.e., the burning of coal, produces approximately 20 to 30 tons of CO_2 per ton of fly ash. Since the worldwide production of Portland cement is expected to reach nearly 2 billion tons by 2010, replacement of any large portion of this cement by fly ash could significantly reduce carbon emissions associated with construction, as long as the comparison takes the production of fly ash as a given.

Embankment

Fly ash properties are unusual among engineering materials. Unlike soils typically used for embankment construction, fly ash has a large uniformity coefficient and it consists of clay-sized particles. Engineering properties that affect the use of fly ash in embankments include grain size distribution, compaction characteristics, shear strength, compressibility, permeability, and frost susceptibility. Nearly all the types of fly ash used in embankments are Class F.

Soil Stabilization

Soil stabilization is the permanent physical and chemical alteration of soils to enhance their physical properties. Stabilization can increase the shear strength of a soil and/or control the shrink-swell properties of a soil, thus improving the load-bearing capacity of a sub-grade to support pavements and foundations. Stabilization can be used to treat a wide range of sub-grade materials from expansive clays to granular materials. Stabilization can be achieved with a variety of chemical additives including lime, fly ash, and Portland cement. Proper design and testing is an important component of any stabilization project. This allows for the establishment of design criteria as well as the determination of the proper chemical additive and admixture rate to be used to achieve the desired engineering properties. Benefits of the stabilization process can include: Higher resistance (R) values, Reduction in plasticity, Lower permeability, Reduction of pavement thickness, Elimination of excavation - material hauling/handling - and base importation, Aids compaction, Provides "all-weather" access onto and within projects sites. Another form of soil treatment closely related to soil stabilization is soil modification, sometimes referred to as "mud drying" or soil conditioning. Although some stabilization inherently occurs in soil modification, the distinction is that soil modification is merely a means to reduce the moisture content of a soil to expedite construction, whereas stabilization can substantially increase the shear strength of a material such that it can be incorporated into the project's structural design. The determining factors associated with soil modification vs soil stabilization may be the existing moisture content, the end use of the soil structure and ultimately the cost benefit provided. Equipment for the stabilization and modification processes include: chemical additive spreaders, soil mixers (reclaimers), portable pneumatic storage containers, water trucks, deep lift compactors, motor graders.

Flowable Fill

Fly ash is also used as a component in the production of flowable fill (also called controlled low strength material, or CLSM), which is used as self-leveling, self-compacting backfill material in lieu of compacted earth or granular fill. The strength of flowable fill mixes can range from 50 to 1,200 lbf/in^2 (0.3 to 8.3 MPa), depending on the design requirements of the project in question. Flowable fill includes mixtures of Portland cement and filler material, and can contain mineral admixtures. Fly ash can replace either the Portland cement or fine aggregate (in most cases, river sand) as a filler material. High fly ash content mixes contain nearly all fly ash, with a small percentage of Portland cement and enough water to make the mix flowable. Low fly ash content mixes contain a high percentage of filler material, and a low percentage of fly ash, Portland cement, and water. Class F fly ash is best suited for high fly ash content mixes, whereas Class C fly ash is almost always used in low fly ash content mixes.

Asphalt Concrete

Asphalt concrete is a composite material consisting of an asphalt binder and mineral aggregate. Both Class F and Class C fly ash can typically be used as a mineral filler to fill the voids and provide contact points between larger aggregate particles in asphalt concrete mixes. This application is used in conjunction, or as a replacement for, other binders (such as Portland cement or hydrated lime). For use in asphalt pavement, the fly ash must meet mineral filler specifications outlined in ASTM D242. The hydrophobic nature of fly ash gives pavements better resistance to stripping. Fly ash has also been shown to increase the stiffness of the asphalt matrix, improving rutting resistance and increasing mix durability.

Geopolymers

More recently, fly ash has been used as a component in geopolymers, where the reactivity of the fly ash glasses is used to generate a binder comparable to a hydrated Portland cement in appearance and properties, but with possibly reduced CO_2 emissions. It should be noted that when the total carbon footprint of the alkali required to form geopolymer cement is considered, including the calcining of limestone as an intermediate to the formation of alkali, the net reduction in total CO_2 emissions may be negligible. Moreover, handling of alkali can be problematic and setting of some geopolymer cements is very rapid (minutes versus hours) as compared to Portland cements, making widespread use of geopolymer cements impractical at the ready mix level.

Roller Compacted Concrete

Another application of using fly ash is in roller compacted concrete dams. Many dams in the US have been constructed with high fly ash contents. Fly ash lowers the heat of hydration allowing thicker placements to occur. Data for these can be found at the US Bureau of Reclamation. This has also been demonstrated in the Ghatghar Dam Project in India.

Bricks

There are several techniques for manufacturing construction bricks from fly ash, producing a wide variety of products. One type of fly ash brick is manufactured by mixing fly ash with an equal amount of clay, then firing in a kiln at about 1000 °C. This approach has the principal benefit of reducing the amount of clay required. Another type of fly ash brick is made by mixing soil, plaster of paris, fly ash and water, and allowing the mixture to dry. Because no heat is required, this technique reduces air pollution. More modern manufacturing processes use a greater proportion of fly ash, and a high pressure manufacturing technique, which produces high strength bricks with environmental benefits.

In the United Kingdom, fly ash has been used for over fifty years to make concrete building blocks. They are widely used for the inner skin of cavity walls. They are naturally more thermally insulating than blocks made with other aggregates.

Ash bricks have been used in house construction in Windhoek, Namibia since the 1970s. There is, however, a problem with the bricks in that they tend to fail or produce unsightly pop-outs. This happens when the bricks come into contact with moisture and a chemical reaction occurs causing the bricks to expand.

In India, fly ash bricks are used for construction. Leading manufacturers use an industrial standard known as "Pulverized fuel ash for lime-Pozzolana mixture" using over 75% post-industrial recycled waste, and a compression process. This produces a strong product with good insulation properties and environmental benefits.

American civil engineer Henry Liu announced the invention of a new type of fly ash brick in 2007. Liu's brick is compressed at 27.58 MPa (272 atm) and cured for 24 hours in a 66 °C steam bath, then toughened with an air entrainment agent, so that it lasts for more than 100 freeze-thaw cycles. Owing to the high concentration of calcium oxide in class C fly ash, the brick can be described as self-cementing. Since this method contains no clay and uses pressure instead of heat, it saves energy, reduces mercury pollution, and costs 20% less than traditional manufacturing techniques. This type of brick is now manufactured under license in the USA.

Some varieties of fly ash brick gain strength as they age.

Metal Matrix Composites

Hollow fly ash can be infiltrated by molten metal to form solid, alumina encased spheres. Fly ash can also be mixed with molten metal and cast to reduce overall weight and density, due to the low density of fly ash. Research is underway to incorporate fly ash into lead acid batteries in a lead calcium tin fly ash composite in an effort to reduce weight of the battery.

Waste Treatment and Stabilization

Fly ash, in view of its alkalinity and water absorption capacity, may be used in combination with other alkaline materials to transform sewage sludge into organic fertilizer or biofuel.

As a Catalyst

Fly ash, when treated with sodium hydroxide, appears to function well as a catalyst for converting polyethylene into substance similar to crude oil in a high-temperature process called pyrolysis.

In addition, fly ash, mainly class C, may be used in the stabilization/ solidification process of hazardous wastes and contaminated soils. For example, the Rhenipal process uses fly ash as an admixture to stabilize sewage sludge and other toxic sludges. This process has been used since 1996 to stabilize large amounts of chromium(VI) contaminated leather sludges in Alcanena, Portugal.

Environmental Problems

Present Production Rate of Fly Ash

In the United States about 131 million tons of fly ash are produced annually by 460 coal-fired power plants. A 2008 industry survey estimated that 43% of this ash is re-used.

Groundwater Contamination

Since coal contains trace levels of arsenic, barium, beryllium, boron, cadmium, chromium, thallium, selenium, molybdenum and mercury, its ash will continue to contain these traces and therefore cannot be dumped or stored where rainwater can leach the metals and move them to aquifers.

Spills of Bulk Storage

Where fly ash is stored in bulk, it is usually stored wet rather than dry so that fugitive dust is minimized. The resulting impoundments (ponds) are typically large and stable for long periods, but any breach of their dams or bunding will be rapid and on a massive scale.

In December 2008 the collapse of an embankment at an impoundment for wet storage of fly ash at the Tennessee Valley Authority's Kingston Fossil Plant resulted in a major release of 5.4 millon cubic yards of coal fly ash, damaging 3 homes and flowing into the Emory River. Cleanup costs may exceed $1.2 billion. This spill was followed a few weeks later by a smaller TVA-plant spill in Alabama, which contaminated Widows Creek and the Tennessee River.

Contaminants

Fly ash contains trace concentrations of heavy metals and other substances that are known to be detrimental to health in sufficient quantities. Potentially toxic trace elements in coal include arsenic, beryllium, cadmium, barium, chromium, copper, lead, mercury, molybdenum, nickel, radium, selenium, thorium, uranium, vanadium, and zinc. Approximately 10% of the mass of coals burned in the United States consists of unburnable mineral material that becomes ash, so the concentration of most trace elements in coal ash is approximately 10 times the concentration in the original coal. A 1997 analysis by the U.S. Geological Survey (USGS) found that fly ash typically contained 10 to 30 ppm of uranium, comparable to the levels found in some granitic rocks, phosphate rock, and black shale.

In 2000, the United States Environmental Protection Agency (EPA) said that coal fly ash did not need to be regulated as a hazardous waste. Studies by the U.S. Geological Survey and others of radioactive elements in coal ash have concluded that fly ash compares with common soils or rocks and should not be the source of alarm. However, community and environmental organisations have documented numerous environmental contamination and damage concerns.

A revised risk assessment approach may change the way coal combustion wastes (CCW) are regulated, according to an August 2007 EPA notice in the Federal Register. In June 2008, the U.S. House of Representatives held an oversight hearing on the Federal government's role in addressing health and environmental risks of fly ash.

Exposure Concerns

Crystalline silica and lime along with toxic chemicals are among the exposure concerns. Although industry has claimed that fly ash is "neither toxic nor poisonous," this is disputed. Exposure to fly ash through skin contact, inhalation of fine particle dust and drinking water may well present health risks. The National Academy of Sciences noted in 2007 that "the presence of high contaminant levels in many CCR (coal combustion residue) leachates may create human health and ecological concerns".

Fine crystalline silica present in fly ash has been linked with lung damage, in particular silicosis. OSHA allows 0.10 mg/m^3, (one ten-thousandth of a gram per cubic metre of air).

Another fly ash component of some concern is lime (CaO). This chemical reacts with water (H_2O) to form calcium hydroxide [$Ca(OH)_2$], giving fly ash a pH somewhere between 10 and 12, a medium to strong base. This can also cause lung damage if present in sufficient quantities.

In a study by NIOSH at a cement company, crystalline silica exposures from fly ash were determined to be of no concern. For maintaining a safe workplace emphasis must be placed on maintaining low nuisance dust levels and to use the appropriate personal protective equipment (PPE). The conclusion of this NIOSH study is supported by other studies devoted to the effects of fly ash on the health of workers in power plants. According to these studies, fly ash dust should not be considered as "silicotic dust", because most of the crystalline silica is coated by amorphic alumino-silicates (glass), the main constituent of fly ash particles. This was shown particularly on the basis of scanning electron microscopy observations, by a research work performed in the Netherlands, but similar findings were obtained also in other countries.

Calcium Sulphate based Building Materials

Calcium sulfate (or calcium sulphate) is a common laboratory and industrial chemical. In the form of γ-anhydrite (the nearly anhydrous form), it is used as a desiccant. It is also used as a coagulant in products like tofu. In the natural state, unrefined calcium sulfate is a translucent, crystalline white rock. When sold as a colour-indicating variant under the name Drierite, it appears blue or pink due to impregnation with Cobalt(II) chloride, which functions as a moisture indicator. The hemihydrate ($CaSO_4 \sim 0.5H_2O$) is better known as plaster of Paris, while the dihydrate ($CaSO_4 \cdot 2H_2O$) occurs naturally as gypsum. The anhydrous form occurs naturally as β-anhydrite. Depending on the method of calcination of calcium sulfate dihydrate, specific hemihydrates are sometimes distinguished: alpha-hemihydrate and beta-hemihydrate. They appear to differ only in crystal shape. Alpha-hemihydrate crystals are more prismatic than beta-hemihydrate crystals and, when mixed with water, form a much stronger and harder superstructure.

Commercial Production and Recovery

The main sources of calcium sulfate are naturally occurring gypsum and anhydrite which occur at many locations worldwide as evaporites. These may be extracted by open-cast quarrying or by deep mining. World production of natural gypsum is around 127 million tonnes per annum.

In addition to natural sources, calcium sulfate is produced as a by-product in a number of processes:

- In flue-gas desulfurization, exhaust gases from fossil-fuel power stations and other processes (e.g. cement manufacture) are scrubbed to reduce their sulfur oxide content, by injecting finely ground limestone or lime. This produces an impure calcium sulfite, which oxidizes on storage to calcium sulfate.
- In the production of phosphoric acid from phosphate rock, calcium phosphate is treated with sulfuric acid and calcium sulfate precipitates.

- In the production of hydrogen fluoride, calcium fluoride is treated with sulfuric acid, precipitating calcium sulfate.
- In the refining of zinc, solutions of zinc sulfate are treated with lime to co-precipitate heavy metals such as barium.
- Calcium sulfate can also be recovered and re-used from scrap drywall at construction sites.

These precipitation processes tend to concentrate radioactive elements in the calcium sulfate product. This is particularly the case with the phosphate by-product, since phosphate rocks naturally contain actinides.

Dehydration Reactions

Heating gypsum to between 100 °C and 150 °C (302 °F) partially dehydrates the mineral by driving off approximately 75% of the water contained in its chemical structure. The temperature and time needed depend on ambient partial pressure of H_2O. Temperatures as high as 170 °C are used in industrial calcination, but at these temperatures γ-anhydrite begins to form. The reaction for the partial dehydration is:

$$CaSO_4{\cdot}2H_2O + \text{heat} \rightarrow CaSO_4{\cdot}\tfrac{1}{2}H_2O + 1\tfrac{1}{2}H_2O\ (\text{steam})$$

The partially dehydrated mineral is called calcium sulfate hemihydrate or calcined gypsum (commonly known as plaster of Paris) ($CaSO_4{\cdot}nH_2O$), where n is in the range 0.5 to 0.8.

The dehydration (specifically known as calcination) begins at approximately 80 °C (176 °F), although in dry air, some dehydration will take place already at 50 °C. The heat energy delivered to the gypsum at this time (the heat of hydration) tends to go into driving off water (as water vapor) rather than increasing the temperature of the mineral, which rises slowly until the water is gone, then increases more rapidly.

The endothermic property of this reaction is exploited by drywall to confer fire resistance to residential and other structures. In a fire, the structure behind a sheet of drywall will remain relatively cool as water is lost from the gypsum, thus preventing (or substantially retarding) damage to the framing (through combustion of wood members or loss of strength of steel at high temperatures) and consequent structural collapse. But at higher temperatures, calcium sulfate will release oxygen and act as an oxidizing agent. This property is used in aluminothermy.

In contrast to most minerals, which when rehydrated simply form liquid or semi-liquid pastes, or remain powdery, calcined gypsum has an unusual property: when mixed with water at normal (ambient) temperatures, it quickly reverts chemically to the preferred dihydrate form, while physically "setting" to form a rigid and relatively strong gypsum crystal lattice:

$$CaSO_4 \cdot ½H_2O + 1½\ H_2O \rightarrow CaSO_4 \cdot 2H_2O$$

This reaction is exothermic and is responsible for the ease with which gypsum can be cast into various shapes including sheets (for drywall), sticks (for blackboard chalk), and molds (to immobilize broken bones, or for metal casting). Mixed with polymers, it has been used as a bone repair cement. Small amounts of calcined gypsum are added to earth to create strong structures directly from cast earth, an alternative to adobe (which loses its strength when wet). The conditions of dehydration can be changed to adjust the porosity of the hemihydrate, resulting in the so-called alpha and beta hemihydrates (which are more or less chemically identical).

On heating to 180 °C, the nearly water-free form, called γ-anhydrite ($CaSO_4 \cdot nH_2O$ where n = 0 to 0.05) is produced. γ-Anhydrite reacts slowly with water to return to the dihydrate state, a property exploited in some commercial desiccants. On heating above 250 °C, the completely anhydrous form called β-anhydrite or "natural" anhydrite is formed. Natural anhydrite does not react with water, even over geological timescales, unless very finely ground.

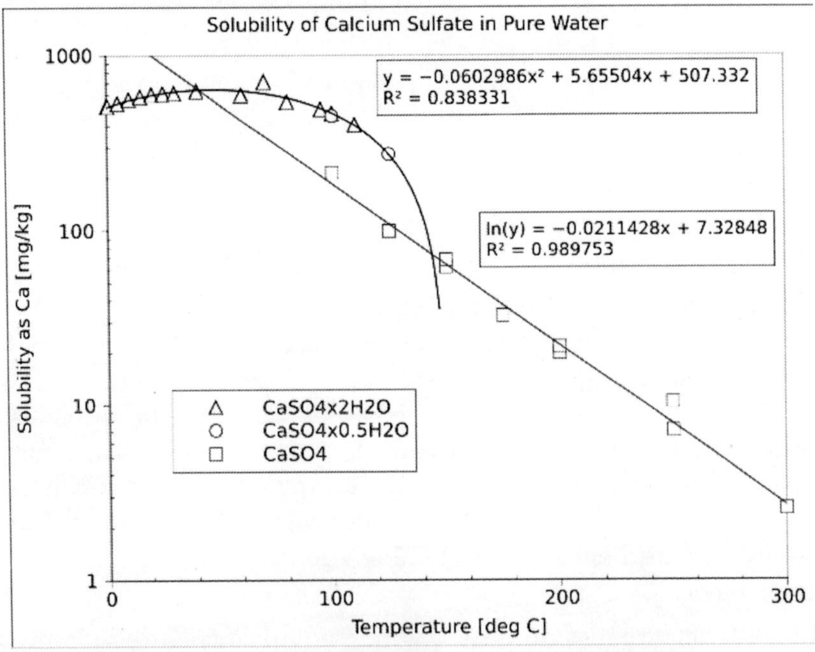

Figure: *Temperature dependence of the solubility of calcium sulfate (3 phases) in pure water.*

The variable composition of the hemihydrate and γ-anhydrite, and their easy inter-conversion, is due to their possessing nearly identical crystal structures, containing "channels" that can accommodate variable amounts of water, or other small molecules such as methanol.

Gypsum

Gypsum is a soft sulfate mineral composed of calcium sulfate dihydrate, with the chemical formula $CaSO_4 \cdot 2H_2O$. It can be used as a fertilizer, is the main constituent in many forms of plaster and is widely mined.

A massive fine-grained white or lightly tinted variety of gypsum, called alabaster, has been used for sculpture by many cultures including Ancient Egypt, Mesopotamia, Ancient Rome, Byzantine empire and the Nottingham alabasters of medieval England. It is the definition of a hardness of 2 on the Mohs scale of mineral hardness. It forms as an evaporite mineral and as a hydration product of anhydrite.

Physical Properties

Gypsum is moderately water-soluble (~2.0–2.5 g/l at 25°C) and, in contrast to most other salts, it exhibits a retrograde solubility, becoming less soluble at higher temperatures. When gypsum is heated in air it loses water and converts first to calcium sulfate hemihydrate, (bassanite, often simply called "plaster") and, if heated further, to anhydrous calcium sulfatc (anhydrite). As for anhydrite, its solubility in saline solutions and in brines is also strongly dependent on NaCl concentration.

Gypsum crystals are found to contain anion water and hydrogen bonding.

Crystal Varieties

Gypsum occurs in nature as flattened and often twinned crystals, and transparent, cleavable masses called selenite. Selenite contains no significant selenium; rather, both substances were named for the ancient Greek word for the Moon.

Selenite may also occur in a silky, fibrous form, in which case it is commonly called "satin spar". Finally, it may also be granular or quite compact. In hand-sized samples, it can be anywhere from transparent to opaque.

A very fine-grained white or lightly tinted variety of gypsum, called alabaster, is prized for ornamental work of various sorts. In arid areas, gypsum can occur in a flower-like form, typically opaque, with embedded sand grains called desert rose. It also forms some of the largest crystals found in nature, up to 12 metres (39 ft) long, in the form of selenite.

Occurrence

Gypsum is a common mineral, with thick and extensive evaporite beds in association with sedimentary rocks. Deposits are known to occur in strata from as far back as the Archaean eon. Gypsum is deposited from lake and sea water, as well as in hot springs, from volcanic vapors, and sulfate solutions in veins. Hydrothermal anhydrite in veins is commonly hydrated to gypsum by groundwater in near-surface exposures. It is often associated with the minerals halite and sulfur. Pure gypsum is white, but other substances found as impurities may give a wide range of colours to local deposits.

Because gypsum dissolves over time in water, gypsum is rarely found in the form of sand. However, the unique conditions of the White Sands National Monument in the US state of New Mexico have created a 710 km^2 (270 sq mi) expanse of white gypsum sand, enough to supply the construction industry with drywall for 1,000 years. Commercial exploitation of the area, strongly opposed by area residents, was permanently prevented in 1933 when president Herbert Hoover declared the gypsum dunes a protected national monument.

Gypsum is also formed as a by-product of sulfide oxidation, amongst others by pyrite oxidation, when the sulfuric acid generated reacts with calcium carbonate. Its presence indicates oxidizing conditions. Under reducing conditions, the sulfates it contains can be reduced back to sulfide by sulfate reducing bacteria. Electric power stations burning coal with flue gas desulfurization produce large quantities of gypsum as a byproduct from the scrubbers.

Orbital pictures from the Mars Reconnaissance Orbiter (MRO) have indicated the existence of gypsum dunes in the northern polar region of Mars, which were later confirmed at ground level by the Mars Exploration Rover (MER) *Opportunity*.

Synthetic gypsum is recovered via flue-gas desulfurization at some coal-fired power plants. It can be used interchangeably with natural gypsum in some applications.

Gypsum also precipitates onto brackish water membranes, a phenomenon known as mineral salt scaling, such as during brackish water desalination of water with high concentrations of calcium and sulfate. Scaling decreases membrane life and productivity. This is one of the main obstacles in brackish water membrane desalination processes, such as reverse osmosis or nanofiltration. Other forms of scaling, such as calcite scaling, depending on the water source, can also be important considerations in distillation, as well as in heat exchangers, where either the salt solubility or concentration can change rapidly.

A new study has suggested that the formation of gypsum starts as tiny crystals of a mineral called bassanite ($CaSO_4 \cdot 0.5H_2O$). This process occurs via a three-stage pathway:

(1) homogeneous nucleation of nanocrystalline bassanite;
(2) self-assembly of bassanite into aggregates, and
(3) transformation of bassanite into gypsum.

Plaster

Plaster is a building material used for coating walls and ceilings. Plaster is manufactured as a dry powder and is mixed with water to form a paste when used. The reaction with water liberates heat through crystallization and the hydrated plaster then hardens. Plaster can be relatively easily worked with metal tools or even sandpaper. These characteristics make plaster suitable for a finishing, rather than a load-bearing material.

The term plaster can refer to gypsum plaster (also known as *plaster of Paris*), lime plaster, or cement plaster.

Types

Gypsum Plaster (Plaster of Paris): Gypsum plaster, or plaster of Paris, is produced by heating gypsum to about 300 °F (150 °C): $CaSO_4 \cdot 2H_2O + heat \rightarrow CaSO_4 \cdot 0.5H_2O + 1.5H_2O$ (released as steam).

When the dry plaster powder is mixed with water, it re-forms into gypsum. The setting of unmodified plaster starts about 10 minutes after mixing and is complete in about 45 minutes; but not fully set for 72 hours. If plaster or gypsum is heated above 392°F (200°C), anhydrite is formed, which will also re-form as gypsum if mixed with water.

A large gypsum deposit at Montmartre in Paris led "calcined gypsum" (roasted gypsum or gypsum plaster) to be commonly known as "plaster of Paris". Plasterers often use gypsum to simulate the appearance of surfaces of wood, stone, or metal, on movie and theatrical sets for example. Nowadays, theatrical plasterers often use expanded polystyrene, although the job title remains unchanged.

Plaster of Paris can be used to impregnate gauze bandages to make a sculpting material called modroc. It is used similarly to clay, as it is easily shaped when wet, yet sets into a resilient and lightweight structure. This is the material that was (and sometimes still is) used to make classic plaster orthopedic casts to protect limbs with broken bones, the medical use having been partly inspired by the artistic use. Set modroc is an early example of a composite material.

Technical Details

Depending on the temperature and duration of the heating process, gypsum converts to the hemihydrate or an anhydrous form. Two polymorphs are known of the hemihydrate. The anhydrous calcium sulfate occurs in three forms, called anhydrite I, II, and III. Each form hydrates differently. The hemihydrate converts to the dihydrate. The anhydrite III converted via the hemihydrate, whereas anhydrite II is converted directly into dihydrate without forming intermediates.

Lime Plaster

Lime plaster is a mixture of calcium hydroxide and sand (or other inert fillers). Carbon dioxide in the atmosphere causes the plaster to set by transforming the calcium hydroxide into calcium carbonate (limestone). Whitewash is based on the same chemistry.

To make lime plaster, limestone (calcium carbonate) is heated to produce quicklime (calcium oxide). Water is then added to produce slaked lime (calcium hydroxide), which is sold as a wet putty or a white powder. Additional water is added to form a paste prior to use. The paste may be stored in air-tight containers. When exposed to the atmosphere, the calcium hydroxide very slowly turns back into calcium carbonate through reaction with atmospheric carbon dioxide, causing the plaster to increase in strength.

Lime plaster was a common building material for wall surfaces in a process known as lath and plaster, whereby a series of wooden strips on a studwork frame was covered with a semi-dry plaster that hardened into a surface. The plaster used in most lath and plaster construction was mainly lime plaster, with a cure time of about a month. To stabilize the lime plaster during curing, small amounts of plaster of Paris were incorporated into the mix. Because plaster of Paris sets quickly, "retardants" were used to slow setting time enough to allow workers to mix large working quantities of lime putty plaster. A modern form of this method uses expanded metal mesh over wood or metal structures, which allows a great freedom of design as it is adaptable to both simple and compound curves. Today this building method has been partly replaced with drywall, also composed mostly of gypsum plaster. In both these methods a primary advantage of the material is that it is resistant to a fire within a room and so can assist in reducing or eliminating structural damage or destruction provided the fire is promptly extinguished.

Lime plaster is used for frescoes, where pigments, diluted in water, are applied to the still wet plaster. USA and Iran are the main plaster producers in the world.

Cement Plaster

Cement plaster is a mixture of suitable plaster, sand, portland cement and water which is normally applied to masonry interiors and exteriors to achieve a smooth surface. Interior surfaces sometimes receive a final layer of gypsum plaster. Walls constructed with stock bricks are normally plastered while face brick walls are not plastered. Various cement-based plasters are also used as proprietary spray fireproofing products. These usually use vermiculite as lightweight aggregate. Heavy versions of such plasters are also in use for exterior fireproofing, to protect LPG vessels, pipe bridges and vessel skirts.

Cement plaster was first introduced in America around 1909 and was often called by the generic name *adamant plaster* after a prominent manufacturer of the time. The advantages of cement plaster noted at that time were its strength, hardness, quick setting time and durability.

Cement Render

Cement rendering is the application of a premixed layer of sand and cement to brick, cement, stone or mud brick. It is often textured, coloured or painted after application. It is generally used on exterior walls but can be used to feature an interior wall. Depending on the 'look' required, rendering can be fine or coarse, textured or smooth, natural or coloured, pigmented or painted.

The cement rendering of brick, concrete and mud houses has been used for centuries to improve the appearance (and sometimes weather resistance) of exterior walls. It can be seen in different forms all over southern Europe. Different countries have their own style and traditional colours. Different finishes can be created by using different tools such as trowels, sponges, or brushes. The art in traditional rendering is, (apart from getting the mix right), the appearance of the top coat. Different tradesmen will have different finishing styles and be able to produce different textures and decorative effects. Some of these special finishing effects may need to be created from a thin finishing 'top coat' or from a finishing wash.

Acrylic Rendering

There is also a wide variety of premixed renders for different situations. Some have a polymer additive to the traditional cement, lime and sand mix for enhanced water resistance, flexibility and adhesion.

Acrylic premixed renders have superior water resistance and strength. They can be used on a wider variety of surfaces, including concrete, cement blocks, and AAC concrete panelling. With the right preparation, they can be used on smoother surfaces like cement sheeting, new high tech polymer exterior cladding such as Uni-Base, and expanded Polystyrene. A few of these require activation with cement just prior to application.

Some of these premixed acrylic renders have a smoother complexion than traditional renders. There is also a wide variety of acrylic bound pigmented 'designer' finishing coats that can be applied over acrylic render. Depending upon the product, they can be rolled, trowelled or sponged on. A limited number can also be sprayed on.

Various finishes, patterns and textures are possible such as sand, sandstone, marble, stone, stone chip, lime wash or clay like finishes. There are stipple, glistening finishes, and those with enhanced water resistance and anti fungal properties. Acrylic renders take only 2 days to dry and cure - much faster than the 28 days for traditional render.

Traditional Rendering

To ensure adhesion, the surface to be rendered is initially hosed off to ensure it is free of any dirt and loose particles. Old paint or old render is scraped away.

The surface is roughened to improve adhesion. For large areas vertical battens are fixed to the wall every 1 to 1.5 metres, to keep the render flat and even.

Cement render consists of 6 parts clean sharp fine sand, 1 part cement and 1 part lime. The lime makes the render more workable and reduces cracking when the render dries. Any general purpose cement can be used. Various additives can be added to the mix for increasing adhesion. Coarser sand is used as the base layer and slightly finer sand for the top layer.

Tremie Concrete

The tremie concrete placement method uses a pipe, through which concrete is placed below water level.

The lower end of the pipe is kept immersed in fresh concrete so that the rising concrete from the bottom displaces the water without washing out the cement content.

Caissons

Concrete is often poured through a tremie pipe in order to build caissons, which are the foundations of, among other things, bridges that span bodies of water.

Monitoring Wells

Tremie methods are employed for materials other than concrete, and for industries other than construction. For example, bentonite slurries for monitoring wells are often emplaced via tremie pipe.

The Greening of Power Plants

Our primary environmental risk this century seems to be the increasing amount of CO2 emission from energy generating processes. The uncontrolled build up of atmospheric CO2 projects directly to catastrophic global warming (over 4o C by century end). Global warming leads to world wide flooding, continental climate change and drought, universal economic depression, regions of famine, mass migration and possibly destruction of the civilized world.

Atmospheric CO2 accumulation will increase as the global economy expands rapidly over the next 10 - 25 years. Can Nanotechnology help to stop the growth or even reduce CO2 emission before it is too late? The preliminary answer is "yes ... use the excess CO2 emissions to make carbon nanotubes ."

A minority of experts disagree about CO2 emission risk. They say that the problem facing the world is an energy shortage, especially in the fossil fuel category required for transportation, and that CO2 accumulation is a secondary issue. One hears "there is a shortage of energy and that will inhibit the global growth of our civilization." That's wrong. A little analysis shows no shortage of energy near term or even longer term. There are over 300 years of coal in the US alone, 50 + years of known recoverable oil reserves worldwide, 75 years of gas currently locked in, an infinite amount of nuclear based power generating potential ... plenty of energy sources without considering the many alternative sources discussed in the trade and professional press.

Because most of the energy sources are not where the energy is used, the unit price to supply energy at a usable location is the guiding variable ... and any alleged shortage of energy is simply resolved by raising that price. Raise energy ($/btu) prices above current market prices, or subsidize currently uneconomic energy generating alternatives, and normal market forces or demand will respond by developing the next least economic energy source on the above list.

For example, raise the price per barrel of sweet crude above $90-100/bbl and the world begins to develop tar sands, shale oil, deep sea oil, alternative solar and wind power, plus serious consideration is given to expanded nuclear power plant construction. There's no shortage.

Just a projected shortage at a particular price. Note that every one of the energy source alternatives produces CO2 that is released to the atmosphere.

Another minority claims that potable water is the primary strategic problem. They say that wars over potable water between growing populations with high lifestyle expectations are inevitable and lead to calamitous sociological disruption. Again, there is obviously plenty of water worldwide. It's called ocean and water can be converted, pumped or purified anywhere if price is no object.

Note that energy is consumed in any proposed potable water solution. If desalinization is a possible solution, large amounts of energy are required to run the desalinization pumps and filters, and to move the refined water to where it may be needed. Projecting out possible world wide need for potable water generation also results in a large need for energy and consequently in increases in CO2 released to the atmosphere. The CO2 problem is ubiquitous.

Many of the residual power plant products can be used productively. Unfortunately that is not true for CO2. CO2, although liquefiable, it is not a natural feedstock in massive quantities. Storage of excess CO2 has focused on underground storage (with pilot plants in various locations) and storage below the deep ocean floor where pressure and temperature make liquid CO2 heavier and denser than water so the ocean's weight and temperature act as a permanent cap on any stored CO2 gas.

Not many experts are comfortable with any of the proposed storage alternatives but investigation continues.

Nanotechnology offers a possible alternative to CO2 storage. Much of the nanotech research with CO2 has concentrated on using CO2 as a catalyst in Carbon Nanotube (CNT) modification. It has not focused on using CO2 simply to manufacture CNTs. Recently, a NY based company that makes machines to manufacture Carbon Nanotubes from various feedstocks casually mentioned to me, as an aside, that they could also manufacture CNT's from CO2 ... that their process could make Carbon Nanotubes from excess Carbon Dioxide. Isn't that interesting? It should be investigated immediately.

If true, one possible solution to the excess CO2 is to locate large conversion machines at power generation facilities to use excess CO2 as a feedstock to make very large quantities of CNT's. These inexpensive CNTs then could be used by society in all the wonderful ways projected for CNT use.

If these auxiliary CNT manufacturing plants were successful, CNT's would be available in quantity at very low cost in very useable volumes many places in the world. Available CNT's would create new nanotech based industries worldwide. It is a very exciting idea.

Since there is an energy deficit in the CO2 to CNT conversion process, some of the power currently being produced by a plant would have to be used to power the conversion of feedstock CO2 to CNT. The additional financial cost could easily be subsidized by governments and institutions because the societal result was so important. Conversion equations built on practical assumptions show that if such a strategy were to be pursued and properly sponsored by all governments with justifiable financial incentives, a significant reduction in CO2 released to the atmosphere would begin occur within 3 - 5 years.

9

Multifunctional and Smart Carbon Nanotube Reinforced Cement Based Materials

A composite cement material is prepared from cement material and carbon nanotubes, wherein the carbon nanotubes are present from about 0.02 wt% to about 0.10 wt % based on weight of cement material. The process for preparing such cement compositions includes sonicating a mixture of a surfactant, water, and carbon nanotubes; and blending the dispersion and the cement material to form a cementitious paste. The process may also include curing the cementitious paste. The composite cement materials are useful in a variety of cement applications where a reduction in nanoscale flaws and fractures is desired.

A significant portion of the current civil infrastructure is partially or completely constructed of cementitious materials, such as concrete. Cementitious materials are typically characterized as quasi-brittle materials with low tensile strength and low strain capacity. Fibres can be incorporated into cementitious matrices to overcome these weaknesses. Typical reinforcement of concrete is provided using reinforcing bars and macrofibres, both of which reinforce concrete on the millimetre scale. Recently, the use of microfibre reinforcement has led to significant improvement of the mechanical properties of cement based materials (Akkaya et al., ACI SP-216-1 2003, 216, 1-18). However, while microfibres delay the development of formed microcracks, they do not stop their initiation. The development of new nanosized fibres has opened a new field for nanosized reinforcement within concrete. The incorporation of fibres at the nanoscale according to the invention will allow the control of the matrix cracks at the nanoscale level and essentially create a new generation of a "crack free material".

One of the most advantageous nanomaterials for nano-reinforcement are carbon nanotubes (CNTs). Since their discovery from lijima, CNTs have opened an incredible range of applications in materials science, electronics, chemical processing, energy management, and many other fields. A CNT can be thought of as a sheet or sheets of graphite that have been rolled up into a tube structure. CNTs can be single walled nanotubes (SWCNTs), as if a single sheet had been rolled up with diameter close to 1 run, or multiwalled (MWCNTs), similar in appearance to a number of sheets rolled together with diameter ranges from 10-80 nm. The unique mechanical, electrical and chemical properties of CNTs make them an attractive candidate for the next generation of composite materials. The Young's modulus of an individual nanotube should be around 1 TPa and its density is about 1.33 g/cm^3. Compared to steel, CNTs possess five to ten times greater modulus than steel at just one sixth the weight. Molecular mechanics simulations suggested that CNTs fracture strains were between 10% and 15% with corresponding tensile stresses on the order of 65 to 93 GPa. The aspect ratios of CNTs are generally beyond 1000.

Carbon nanotubes are expected to have several distinct advantages as a reinforcing material for cement as compared to more traditional fibres. First, they have significant greater strengths than conventional fibres, which strengths should improve overall mechanical behavior. Second, they exhibit higher aspect ratios, requiring significantly higher energies for crack propagation than would be the case for a lower aspect ratio fibre. Thirdly, CNTs exhibit smaller diameters which means that, provided that they are uniformly dispersed according to the invention, they can be widely distributed in the cement matrix with reduced fibre spacing. The tensile strength of the composite is increased when the fibre free area is decreased. However, the potential of using nanotubes as reinforcement for cementitious materials has not been fully realised heretofore mainly because of the difficulties in processing. The two major drawbacks associated with the incorporation of CNTs in cement based materials are poor dispersion and cost. CNTs tend to adhere together due to Van der Waal forces, and it is particularly difficult to separate them individually. To achieve good reinforcement in a composite, it is critical to have uniform dispersion of CNTs within the matrix. Poor dispersion of CNTs leads to the formation of many defect sites in the nanocomposite and limits the efficiency of the CNTs in the matrix

Earlier attempts have been made to add CNTs in cementitious matrices at an amount ranging from 0.5 to 2.0 wt% (by weight of cement). Prior work on CNTs in liquid dispersions has focused on pre-treatment of the nanotube"s surface via chemical modification.

The present invention provides a solution to the above-stated dispersion and cost problems by providing improved dispersion of small amounts of carbon nanotubes in cement based materials. In an embodiment of the invention, the method involves applying ultrasonic energy and use of a surfactant to provide a CNT fluid (e.g. aqueous) dispersion and mixing the CNT dispersion with a cement material in a manner that the CNTs are well dispersed in the cementitious matrix. As a result of effective dispersion, the fracture properties and other properties such as the transport properties of the cementitious matrix are substantially increased by adding a very low, less costly amount of CNTs.

The addition of a small quantity of CNTs enables the control of the matrix cracks at the nanoscale level. In addition the use of CNTs at a very low percentage makes the cost of the cement composition pursuant to the invention very attractive. In an illustrative embodiment of the invention, the CNT' s are present in an amount from about 0.02% to about 0.10%, preferably about 0.03% to about 0.10%, by weight of cement.

The nanocomposite cementitious material pursuant to the invention will have a beneficial impact on economics, environment, and structural durability, which is extremely important to the overall public view. The potential benefits of infrastructure having such a new type of composite material include the following:

- Enhance the quality of cement and improve material durability by delaying the crack formation and increase the likelihood of multiple cracking.
- Reduce the maintenance construction cost.
- Develop a high performance and durable repair material which will produce durable, aesthetically pleasing and cost-effective repairs.

The CNT-cement nanocomposite material pursuant to the invention will find wide application for highway structures, bridges, pavements, runways for airports, continuous slab-type sleepers for high speed trains and in general in all applications of conventional and high strength concrete, as well as in manufactured precast elements for residential and commercial buildings.

Applications of Nanotechnology in Road Pavement Engineering

Pavement engineering is a branch of civil engineering that uses engineering techniques to design and maintain flexible (asphalt) and rigid (concrete) pavements. This includes streets and highways and involves knowledge of soils, hydraulics, and material properties. Pavement engineering involves new construction as well as rehabilitation and maintenance of existing pavements. Maintenance often involves using engineering judgment to make maintenance repairs with the highest long-term benefit and lowest cost. The Pavement Condition Index (PCI) is an example of an engineering approach applied to existing pavements. Another example is the use of a falling weight deflectometre (FWD) to non-destructively test existing pavements. Calculation of pavement layer strengths can be performed from the resulting deflection data.

Application of Nanoscience Modelling to Understand the Atomic Structure of CSH

Calcium silicate hydrate (C-S-H) is a main product of Portland cement hydration. Although C-S-H is treated as an amorphous phase in macro scale, the atomic-scale structure of C-S-H is considered to be nanocrystalline and short-ordered. Various studies have been carried out and many models have been proposed in the past few decades to describe the structure of C-S-H at nano scale. Tobermorite and jennite are considered as two naturally existing crystalline solids sharing similar structures with C-S-H, especially tobermorite. As a result, tobermorite-based models are predominant in the literature. Besides, combination models with both tobermorite and jennite as well as exceptive models have also been developed in the same period.

Preparation

Calcium silicate hydrate (also shown as C-S-H) is a result of the reaction between the silicate phases of Portland cement and water. This reaction typically is expressed as:

$$2\ Ca_3SiO_5 + 7\ H_2O \rightarrow 3\ CaO \cdot 2\ SiO_2 \cdot 4\ H_2O + 3\ Ca(OH)_2 + 173.6\ kJ$$

The stoichiometry of C-S-H in cement paste is variable and the state of chemically and physically bound water in its structure is not transparent, which is why "-" is used between C, S, and H.

Synthetic C-S-H can be prepared from the reaction of CaO and SiO_2 in water or through the double precipitation method using various salts. These methods provide the flexibility of producing C-S-H at specific C/S ratios. The C-S-H from cement phases can also be treated with ammonium nitrate in order to achieve desired C/S ratio.

Properties

The crystal structure of C-S-H in cement paste has not been fully resolved yet and there is still ongoing debate over its nanostructure.

The SEM micrographs of C-S-H does not show any specific crystalline form. They usually manifest as flakes or dendrites/fibrils.

Synthetic C-S-H can be divided in two categories separated at the C/S ratio of about 1.1. There are several indications that the chemical, physical and mechanical characteristics of C-S-H varies noticeably between these two categories. The Effect of SWCNT and Other Nanomaterials on Cement Hydration and Reinforcement

Alkali–Silica Reaction

The alkali–silica reaction (ASR) is a reaction which occurs over time in concrete between the highly alkaline cement paste and reactive non-crystalline (amorphous) silica, which is found in many common aggregates. The ASR reaction is the same as the pozzolanic reaction, which is a simple acid-base reaction between calcium hydroxide, also known as Portlandite, or ($Ca(OH)_2$), and silicic acid (H_4SiO_4, or $Si(OH)_4$). For the sake of simplicity, this reaction can be schematically represented as following:

$$Ca(OH)_2 + H_4SiO_4 \rightarrow Ca^{2+} + H_2SiO_4^{2-} + 2\ H_2O \rightarrow CaH_2SiO_4 \cdot 2\ H_2O$$

This reaction causes the expansion of the altered aggregate by the formation of a swelling gel of calcium silicate hydrate (C-S-H). This gel increases in volume with water and exerts an expansive pressure inside the material, causing spalling and loss of strength of the concrete, finally leading to its failure.

ASR can cause serious expansion and cracking in concrete, resulting in critical structural problems that can even force the demolition of a particular structure.

The mechanism of ASR causing the deterioration of concrete can be described in four steps as follows:

1. The alkaline solution attacks the siliceous aggregate, converting it to viscous alkali silicate gel.
2. Consumption of alkali by the reaction induces the dissolution of Ca^{2+} ions into the cement pore water. Calcium ions then react with the gel to convert it to hard C-S-H.
3. The penetrated alkaline solution converts the remaining siliceous minerals into bulky alkali silicate gel. The resultant expansive pressure is stored in the aggregate.

4. The accumulated pressure cracks the aggregate and the surrounding cement paste when the pressure exceeds the tolerance of the aggregate.

ASR can be mitigated in new concrete by three complementary approaches:

1. *Limit the alkali metal content of the cement.* Many standards impose limits on the "Equivalent Na_2O" content of cement.
2. *Limit the reactive silica content of the aggregate.* Certain volcanic rocks are particularly susceptible to ASR because they contain volcanic glass (obsidian) and should not be used as aggregate. The use of calcium carbonate aggregates is sometimes envisaged as an ultimate solution to avoid any problem. However, while it may be considered as a necessary condition, it is not a sufficient one. In principle, limestone ($CaCO_3$) is not expected to contain a high level of silica, but it actually depends on its purity. Indeed, some siliceous limestones (a.o., *Kieselkalk* found in Switzerland) may be cemented by amorphous or poorly crystalline silica and can be very sensitive to the ASR reaction, as observed with some siliceous limestones exploited in quarries in the area of Tournai in Belgium. So, the use of limestone as aggregate is not a guarantee against ASR in itself. The silica content of the limestone and its reactivity must remain below a threshold value that has to be carefully experimentally assessed by the aggregate producer.
3. Add very fine siliceous materials to neutralize the excessive alkalinity of cement with silicic acid by voluntary provoking a controlled pozzolanic reaction at the early stage of the cement setting. Convenient pozzolanic materials to add to the mix may be, *e.g.*, pozzolan, silica fume, fly ash, or metakaolin. These react preferentially with the cement alkalis without formation of an expansive pressure, because siliceous minerals in fine particles convert to alkali silicate and then to calcium silicate without formation of semipermeable reaction rims.

In other words, as it is sometimes possible *to fight fire with fire*, it is also feasible to combat the ASR reaction by itself. A prompt reaction initiated at the early stage of concrete hardening on very fine silica particles will help to suppress a slow and delayed reaction with large siliceous aggregates on the long term. Following the same principle, the fabrication of low-pH cement also implies the addition of finely divided pozzolanic materials rich in silicic acid to the concrete mix to decrease its alkalinity.

As part of a study conducted by the Federal Highway Administration, a variety of methods have been applied to field structures suffering from ASR-affected expansion and cracking. Some methods, such as the application of silanes, have shown significant promise, especially when applied to elements such as small columns and highway barriers, whereas other methods, such as the topical application of lithium compounds, have shown little or no promise in reducing ASR-induced expansion and cracking.

ASR Test

Some ASTM Tests that screen aggregate for the potential of ASR include:

ASTM C 227: "Test Method for Potential Alkali Reactivity of Cement-Aggregate Combinations (Mortar-Bar Method)"

ASTM C289: Standard Test Method for Potential Alkali-Silica Reactivity of Aggregates (Chemical Method)

ASTM C 295: "Guide for Petrographic Examination of Aggregate for Concrete"

ASTM C1260: "Test Method for Potential Reactivity of Aggregates (Mortar-Bar-Test)"

ASTM C1293: "Test Method for Concrete Aggregates by Determination of Length Change of Concrete Due to Alkali-Silica Reaction"

ASTM C1567: Standard Test Method for Determining the Potential Alkali-Silica Reactivity of Combinations of Cementitious Materials and Aggregate (Accelerated Mortar-Bar Method)

The concrete microbar test was proposed by Grattan-Bellew *et al.* (2003) as a universal accelerated test for alkali-aggregate reaction.

Alkali–Aggregate Reaction

Alkali–aggregate reaction is a term mainly referring to a reaction which occurs over time in concrete between the highly alkaline cement paste and non-crystalline silicon dioxide, which is found in many common aggregates. This reaction can cause expansion of the altered aggregate, leading to spalling and loss of strength of the concrete.

More Accurate Terminology

The alkali–aggregate reaction is a general, but relatively vague, expression which can lead to confusion. More precise definitions include the following:

1. Alkali–silica reaction (ASR, the most common reaction of this type);
2. Alkali–silicate reaction, and;
3. Alkali–carbonate reaction.

The alkali–silica reaction is the most common form of alkali–aggregate reaction.

Two other types are:

- the alkali–silicate reaction, in which layer silicate minerals (clay minerals), sometimes present as impurities, are attacked, and ;
- the alkali–carbonate reaction, which is an uncommon attack on certain argillaceous dolomitic limestones, likely involving the expansion of the mineral Brucite ($Mg(OH)_2$.

The Pozzolanic reaction which occurs in the setting of the mixture of slaked lime and pozzolanic materials has also features similar to the alkali–silica reaction, mainly the formation of calcium silicate hydrate (CSH).

Optimization of Clay Addition for the Enhancement of Pozzolanic Reaction in Nanomodified Cement Paste

A structure is considered smart when it is made from a material that, among other things, has certain desirable properties (multi-functionality). Such properties, when monitored non-destructively, can provide information regarding the integrity during the lifetime of a structure (smart). The real-time monitoring of the multifunctional characteristics of the material enables the prediction of the level of damage and/or the potential failure of the structure. Smart cement-based structures should be therefore made out of modified concrete with properties different than those of the conventional concrete. The conventional concrete is not electrically conductive; its heat capacity is high, while its thermal insulation property is low. In addition, its flexural strength and toughness are low, therefore requiring steel reinforcement in large quantities for using it in structural applications. Target properties of a cement-based multifunctional material should be: exceptional mechanical properties which would enable a highly resistant lightweight structure, electrical and thermal conductivity, and piezo-electric characteristics.

Pozzolan

A pozzolan is an siliceous or siliceous and aluminous material which, in itself, possesses little or no cementitious value but which will, in finely divided form and in the presence of water, react chemically with calcium hydroxide at ordinary temperature to form compounds possessing cementitious properties (ASTM C618). The broad definition of a pozzolan imparts no bearing on the origin of the material, only on its capability of reacting with calcium hydroxide and water. A quantification of this capability is comprised in the term pozzolanic activity.

Pozzolanic Materials

The general definition of a pozzolan embraces a large number of materials which vary widely in terms of origin, composition and properties. Both natural and artificial (man-made) materials show pozzolanic activity and are used as supplementary cementitious materials. Artificial pozzolans can be produced deliberately, for instance by thermal activation of kaolin-clays to obtain metakaolin, or can be obtained as waste or by-products from high-temperature process such as fly ashes from coal-fired electricity production. The most commonly used pozzolans today are industrial by-products such as fly ash, silica fume from silicon smelting, highly reactive metakaolin, and burned organic matter residues rich in silica such as rice husk ash. Their use has been firmly established and regulated in many countries. However, the supply of high-quality pozzolanic by-products is limited and many local sources are already fully exploited. Alternatives to the established pozzolanic by-products are to be found on the one hand in an expansion of the range of industrial by-products or societal waste considered and on the other hand in an increased usage of naturally occurring pozzolans.

Natural pozzolans are abundant in certain locations and are extensively used as an addition to Portland cement in countries such as for example Italy, Germany, Greece and China. The great majority of natural pozzolans in use today is of volcanic origin. Volcanic ashes and pumices largely composed of volcanic glass are commonly used, as are deposits in which the volcanic glass has been altered to zeolites by interaction with alkaline waters. Deposits of sedimentary origin are less common. Diatomaceous earths, formed by the accumulation of siliceous diatom microskeletons, are a prominent source material here.

Use

The benefits of pozzolan utilisation in cement and concrete are threefold. First is the economic gain obtained by replacing a substantial part of the Portland cement by cheaper, pollution free, natural pozzolans or industrial by-products. Second is the lowering of the blended cement environmental cost associated with the greenhouse gases emitted during Portland cement production. A third advantage is the increased durability of the end product. Additionally, the increased blending of pozzolans with Portland cement is of limited interference in the conventional production process and offers the opportunity to valorize large amounts of industrial and societal waste into durable construction materials.

Current practice may permit up to a 40 percent reduction of Portland cement used in the concrete mix when replaced with a carefully designed combination of approved pozzolans.

When the concrete mix proportions of sand :pozzalan: gravel: water:cement is chosen, concrete can utilise pozzolans to ensure setting, increase durability, reduce cost and reduce pollution without significantly reducing the final compressive strength or other performance characteristics.

The properties of hardened blended cements are strongly related to the development of the binder microstructure, i.e., to the distribution, type, shape and dimensions of both reaction products and pores. The beneficial effects of pozzolan addition in terms of higher compressive strength performance and greater durability are mostly attributed to the pozzolanic reaction in which calcium hydroxide is consumed to produce additional C-S-H and C-A-H reaction products. These pozzolanic reaction products fill in pores and result in a refining of the pore size distribution or pore structure. This results in a lowered permeability of the binder.

The contribution of the pozzolanic reaction to cement strength is usually developed at later curing stages, depending on the pozzolanic activity. In the large majority of blended cements initial lower strengths can be observed compared to the parent Portland cement. However, especially in the case of pozzolans finer than the Portland cement, the decrease in early strength is usually less than what can be expected based on the dilution factor. This can be explained by the filler effect, in which small SCM grains fill in the space between the cement particles, resulting in a much denser binder. The acceleration of the Portland cement hydration reactions can also partially accommodate the loss of early strength.

The increased chemical resistance to the ingress and harmful action of aggressive solutions constitutes one of the main advantages of pozzolan blended cements. The improved durability of the pozzolan-blended binders enables to lengthen the service life of structures and reduces the costly and inconvenient need to replace damaged constructions. One of the principal reasons increased durability is the lowered calcium hydroxide content available to take part in deleterious expansive reactions induced by e.g. sulfate attack. Furthermore, the reduced binder permeability slows down the ingress of harmful ions such as chlorine or carbonate. The pozzolanic reaction can also reduce the risk of expansive alkali-silica reactions between the cement and aggregates by changing the binder pore solution. Lowering the solution alkalinity and increasing alumina concentrations strongly decreases or inhibits the dissolution of the aggregate aluminosilicates.

Pozzolanic Reaction

The pozzolanic reaction is the chemical reaction that occurs in portland cement containing pozzolans. It is the main reaction involved in the Roman concrete invented in Ancient Rome and used to build, for example, the Pantheon. At the basis of the pozzolanic reaction stands a simple acid-base reaction between calcium hydroxide, also known as Portlandite, or ($Ca(OH)_2$), and silicic acid (H_4SiO_4, or $Si(OH)_4$), and . Simply, this reaction can be schematically represented as follows:

$$Ca(OH)_2 + H_4SiO_4 \rightarrow Ca^{2+} + H_2SiO_4^{2-} + 2\ H_2O \rightarrow CaH_2SiO_4 \cdot 2\ H_2O$$

or summarized in abbreviated notation of cement chemists:

$$CH + SH \rightarrow C\text{-}S\text{-}H$$

The product of general formula ($CaH_2SiO_4 \cdot 2\ H_2O$) formed is a calcium silicate hydrate, also abbreviated as C-S-H in cement chemist notation, the hyphenation denotes the variable stoichiometry. The ratio Ca/Si, or C/S, and the number of water molecules can vary and the above mentioned stoichiometry may differ.

Many pozzolans contain aluminate, or $Al(OH)_4^-$, that will react with calcium hydroxide and water to form calcium aluminate hydrates such as C_4AH_{13}, C_3AH_6 or hydrogarnet, or in combination with silica C_2ASH_8 or strätlingite (cement chemist notation). In the presence of anionic groups such as sulphate, carbonate or chlorine, AFm phases and AFt or ettringite phases can form.

Curing, Hydration, and Microstructure of Cement Paste

Proper curing of concrete is widely recognised as a necessity for assuring adequate field performance of concrete structures.

Not only is it recognised as important to minimize evaporation of the concrete mixture water, but it is equally emphasized to provide a source of external (or internal) curing water to replace that consumed by chemical shrinkage during the hydration of the cement. A cement paste, mortar, or concrete cured under sealed conditions will self-desiccate, resulting in the creation of coarse capillary pores within the microstructure.

For water-cement ratios (w/c) greater than approximately 0.42, there is sufficient water in the mixture such that complete hydration of the cement can be achieved theoretically without supplying additional water to the cement paste.

Even if complete hydration were achievable, however, the lack of additional curing water

may still result in the creation of relatively large pores within the final microstructure. In this case, the addition of curing water would assure that all pores remain water-filled and eligible as locations for the precipitation and growth of hydration products during curing.

Proper curing of the newer high-performance concretes is both more critical and more difficult.

Thus, high-performance concrete has been plagued by ubiquitous early-age cracking problems, and novel solutions, such as the use of internal curing, 3-5 have been developed to avoid or at least reduce such cracking. Early-age cracking is a complex phenomenon, depending on thermally induced strains, autogenous stresses, and strains due to self-desiccation and development of mechanical properties, among others.

The microstructure of hydrating cement paste within concrete clearly has a large influence on this phenomenon. Thus, a more fundamental understanding of how this microstructure develops during the course of early hydration, as a function of both w/c an curing conditions, would assist in optimizing current curing strategies and also in formul ating new and innovative ones.

The concurrent goals of hydration are to connect the original cement particles into as strong a network as possible and to disconnect the original water-filled capillary pore space as much as possible. Understanding the influence of curing practices on this microstructure development is critical to developing rational curing practices and integrating curing into the mixture design process.

General Features

The hydration products in cement paste do not exist in isolation, but are entangled with one another at the scale of microns and even nanometres. The possible exception to this is the high-density C-S-H phase that forms within the boundaries of the original cement grains. Thus in describing the microstructure of cement paste the structure of the individual solid phases is not as relevant as the overall distribution of solid phases and porosity. With this in mind, a basic but useful description of cement paste microstructure is comprised of just three "phases"

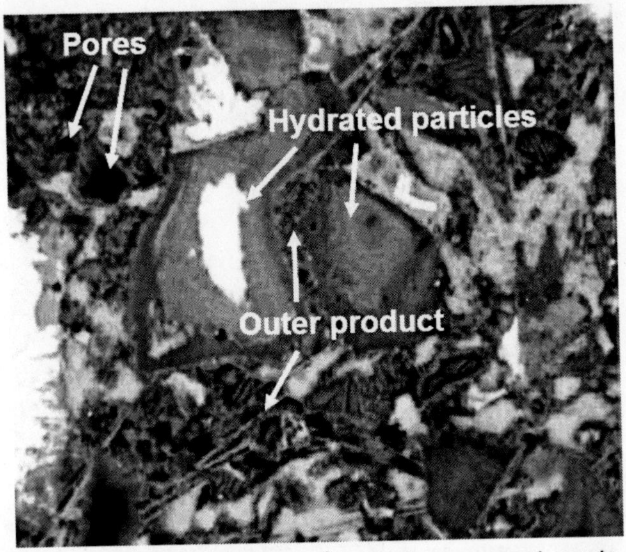

Figure: Backscattered SEM image of a mature cement paste showing the main microstructural features.

Hydrated cement particles: These consist of high-density C-S-H, and in some cases an interior core of unhydrated cement. These behave as individual solid particles within a continuous matrix analogous to the aggregate particles in concrete. These features are sometimes called "phenograins", which simply refers to the fact that they are distinctly visible in a microscope.

Outer hydration product: This is the continuous phase that grows within the capillary pore space and binds the cement together. Following the analogy used above, it plays a role similar to cement paste in concrete. It consists of solid C-S-H gel, gel pores, calcium hydroxide, and calcium sulfoaluminate phases. This "phase" appears as various shades of speckled grey in an optical or electron microscope, and is sometimes referred to as "groundmass." By far the most important individual phase is, of course, the low density C-S-H gel (and its gel pores), because its high surface area gives this phase its strength.

Large pores: These consist of true capillary pores, entrapped air voids, and the entrained air system. These features appear as discrete black voids in a microscope. The large pore system can be continuous or discontinuous, depending on the degree of hydration and starting w/c, but this cannot be determined from microscopy.

Autoclaved Aerated Concrete

Autoclaved aerated concrete (AAC), also known as autoclaved cellular concrete (ACC), autoclaved lightweight concrete (ALC), autoclaved concrete, cellular concrete, porous concrete, Aircrete, Hebel Block, and Ytong is a lightweight, precast, concrete building material invented in the mid-1920s that simultaneously provides structure, insulation, and fire- and mold-resistance. AAC products include blocks, wall panels, floor and roof panels, cladding (facade) panels and lintels.

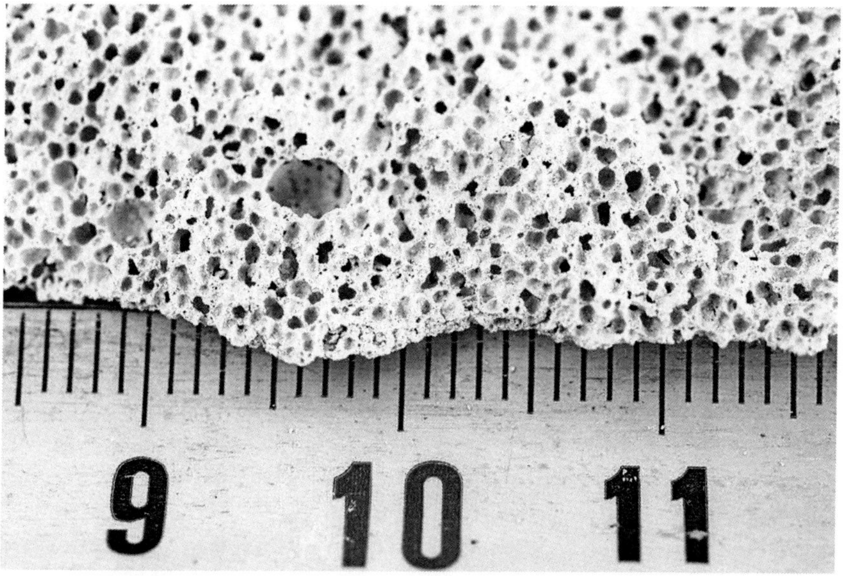

Figure: *Aerated autoclaved concrete - close-up view*

Figure: *Autoclaved aerated concrete tiles of various shapes and sizes*

AAC was perfected in the mid-1920s by the Swedish architect and inventer Dr. Johan Axel Eriksson, working with Professor Henrik Kreüger at the Royal Institute of Technology. It went into production in Sweden in 1929 in a factory in Hällabrottet and quickly became very popular. In the 1940s, the trademark *Ytong* was introduced, and was often referred to as "blue concrete" in Sweden due to its blueish tinge. This version of Ytong was produced from alum shale, whose combustible carbon content made it beneficial to use in the production process. The competing concrete brand *Siporex* produced in Tuzla, Bosnia, used other raw materials. "Ytong" acquired *Siporex* and today produces "Siporex" under "Ytong" brand in Tuzla, Bosnia factory. Unfortunately, the slate deposits used for Ytong also contain uranium, which makes the material give off radioactive radon gas to the building. In 1972, the Swedish Radiation Safety Authority pointed out the unsuitability of a radon-emitting construction material, and the use of alum slate in the production of Ytong ceased in 1975. Ytong produced after 1975 has used raw materials without the uranium content.

AAC production in Europe has slowed down considerably, but the industry is growing rapidly in Asia due to strong demand in housing and commercial space. China, Central Asia, India, and the Middle-East are the biggest markets in terms of AAC manufacturing and consumption.

Bibliography

Altmann, J. & Gubrud, M.A.: *Discovering the Nanoscale*, IOS Press, Amsterdam, 2004.

Andrew, N.: *Foundations of Nanomechanics: from Solid-state Theory to Device Applications*, Berlin, New York, 2003.

Attebery, B.: *Nanoculture: Implications of the New Technoscience*, Intellect Books, Bristol, 2004.

Bainbridge, W.S. : *Societal Implications of Nanoscience and Nanotechnology*, Dordrecht, Kluwer, 2001.

Bast, R.C. Jr.: *Cancer Medicine*, Decker, Hamilton, 2000.

Berube, D.M.: *Discovering the Nanoscale*, IOS Press, Amsterdam, 2004.

Brosnan, J.: *The Primal Scream: A History of Science Fiction Film*, Orbit Books, London, 1991.

Carroll, J.S.: *Social Science Research Methods for Assessing Societal Implications of Nanotechnology*, Dordrecht, Kluwer, 2001.

Caviedes, C.: *El Nino in History: Storming Through the Ages*, University Press of Florida, Florida, 2001.

Credi, A.: *Molecular Devices and Machines: a Journey into the Nano World*, Wiley-VCH, Weinheim, 2003.

Crow, M. & Sarewitz, D.: *Societal Implications of Nanoscience and Nanotechnology*, Dordrecht: Kluwer, 2001.

Daniel Ratner: *Nanotechnology: a Gentle Introduction to the Next Big Idea*, Upper Saddle River, Prentice Hall, 2003.

Dobzhanshy, T.: *Genetics of the Evolutionary Process*, Columbia University Press, New York, 1970.

Drexler, K. Eric: *Engines of Creation: The Coming Era of Nanotechnology*, Anchor Books, New York, 1986.

Dyson, Tim: *India's Historical Demography: Studies in Famine, Disease, and Society*, Riverdale, Westwood, MA, 1989.

Eamonn Gearon: *The Sahara: A Cultural History*, Signal Books, UK, 2011.

Frank J. Owens: *Introduction to Nanotechnology*, Hoboken, J. Wiley, 2003.

Glimell, H.: *Dynamics of the Emerging Field of Nanoscience*, Dordrecht, Kluwer, 2001.

Glimell, H.: *Societal Implications of Nanoscience and Nanotechnology*, Kluwer, Dordrecht, 2001.

Haber, L. F.: *The Chemical Industry: 1900-1930, International Growth and Technological Change*, Oxford: Claredon Press, 1971.

Hessenbruch, A.: *Nanotechnology and the Negotiation of Novelty*, Amsterdam, IOS Press, 2004.

James Lewis: *Prospects in Nanotechnology: Toward Molecular Manufacturing,* John Wiley & Sons, New York, 1995.

Karn, Barbara: *Nanotechnology and the Environment: Applications and Implications,* American Chemical Society, Washington D.C., 2004.

Keblinski, P. : *Encyclopedia of Nanoscience and Nanotechnology*, American Scientific, Los Angeles, Calif, USA, 2004.

Liu W.K., Karpov E.G., Park H.S.: *Nano Mechanics and Materials,* Wiley, Weinheim, 2006.

Mayr, E.: *Change of Genetic Environment and Evolution*, Allen and Unwin, London, 1954.

Mody, C.C.M.: *How Probe Microscopists Became Nanotechnologists, Discovering the Nanoscale*, IOS Press, Amsterdam, 2004.

Norio Taniguchi: *Nanotechnology: Integrated Processing Systems for Ultra-precision and Ultra-fine Products,* Oxford University Press, New York, 1996.

Orth, R. N.: *Nanometer Scale Antibody Patterning for Directed cell Immobilization and Stimulation,* Cornell University, Ithaca, 2002.

Pressman, J.: *Nano Narrative: A Parable from Electronic Literature*, Intellect Books, Bristol, 2004.

Pressman, J.: *Nanoculture: Implications of the New Technoscience*, Intellect Books, Bristol, UK, 2004.

Ratner, Daniel: *Nanotechnology: a Gentle Introduction to the Next Big Idea*, Prentice Hall, Upper Saddle River, 2003.

Robison, W.L.: *Discovering the Nanoscale*, IOS Press, Amsterdam, 2004.

Sarewitz, D.: *Nanotechnology and Societal Transformation*, Dordrecht, Kluwer, 2001.

Stephen M. Goodnick: *Transport in Nanostructures,* Cambridge University Press, Cambridge, 1997.

Tenner, E.: *Nanotechnology and Unintended Consequences*, , Dordrecht, Kluwer, 2001.

Vasudev Bhatnagar: *Cell Science and Technology*, Campus Books, Delhi, 2009.

Vermaas, P.E.: *Discovering the Nanoscale*, IOS Press, Amsterdam, 2004.

Weil, V.: *Ethical Issues in Nanotechnology*, Dordrecht, Kluwer, 2001.

Index

❑❑❑